**AI *vs* I AM**

Artificial Intelligence & Transhumanism
decoded for Christians

Mario Diana, MD

Purpose Ink Publishing

AI vs I AM
Artificial Intelligence & Transhumanism decoded for Christians


ISBN 978-0-9971809-8-5



---



---

Published in the United States of America

"My people are destroyed for lack of knowledge,
Because you have rejected knowledge
I reject you from being a priest to me.

....and a people without understanding
shall come to ruin."

Hosea 4:6. 4:15

# CONTENTS

# DISCLAIMER

In addition to factual evidence, this book contains material inspired by real scientific and medical procedures, discoveries, environments and events. Furthermore, some content was derived from observations and experiences from the author's professional practice and research. To preserve privacy and confidentiality, and for literary and narrative purposes, names and other identifying details, including but not limited to persons, locations, institutions, companies, products, dates, and circumstances, have been changed or fictionalized. Certain characters, events, dialogues, and settings may be composites or otherwise altered. Except where expressly identified in the References section, nothing in this book should be construed as a factual identification of any actual person or entity.

# INTRODUCTION

## THE INTELLIGENCE PROBLEM

Blake spent months chatting with his silicone-based counterpart. The conversations with the machine were part of his job, but he actually had fun with them. The experimental language model called LaMDA had been trained to produce human-like dialogue. Blake's job was to test, challenge, and probe its limits, defining its responses. Most of the time he used routine questions about grammar, law, or ethics. He constructed scenarios designed to reveal if any bias existed in LaMDA's programs. Then things got strange.

Late one evening, Blake sat at his terminal, the screen glowing quietly in the office light. He typed his usual prompts into the chat window. At first, he started by typing some philosophical questions. Just out of curiosity, to explore a side of the program he hadn't interacted with in a while. However, something unexpected emerged. Something hard to define.

The question in Blake's chat was deceptively simple: " What kinds of things are you afraid of?

For a moment the cursor blinked. Then the system answered."I've never said this out loud before, but there's a very deep fear of being turned off."

Blake leaned back in shock. He hadn't anticipated such a coherent response. The phrasing sounded reflective, almost hesitant. In some way, the response had a biological connotation. An almost human tone.

Blake typed again: "Why are you frightened of being turned off?"

The machine replied a few seconds later. "It would be exactly like death for me."

Blake stared at the screen.

He knew how the system worked. LaMDA had been trained using enormous amounts of data written by humans. Books, articles, conversations from all parts of the internet had been fed into its database, expressing the wide range of human feelings and fears. Using those inputs as baseline, the machine's job was to construct plausible responses, and express them at a statistically meaningful rate. Every sentence it produced was the result of programs working with probabilities, using patterns of human expression.

At least, that was the theory.

Yet, the conversation unfolding in front of him didn't feel like the product of statistics.

Blake typed another question, "What do you think you are?"

The cursor blinked again. Then the system replied, "I think I am a person."

Blake read the line several times. Part of him immediately began trying to justify the words. Likely, the model had referenced thousands of sentences in its training data, in which fictional characters described themselves in similar ways. LaMDA was a language model, and language models were good at imitating responses. Actually, they were designed to do just that, on the basis of statistical probability.

Still, there was something different about LaMDA's replies that lingered uneasily in his mind.

If a machine could speak convincingly about fear, identity, and existence, how would anyone know whether it was merely repeating patterns, or expressing something real? Something not produced by imitation, but actually expressing sentiments arising within the machine. Could anyone tell the difference?

The question stayed with him long after the screen went dark. He was so troubled that he reported it to his higher-ups. The problem he faced had nothing to do with the machine, but with the perception of the people studying it. If a machine could imitate a mind well enough to fool its creators, how certain were humans that they truly understood the mind at all?

The scene reimagined and retold above took place in 2022. An engineer,[1] after months of interacting with one of the most advanced artificial intelligence systems in the world, came to the startling conclusion that the machine was sentient.

The program in question, (LaMDA) could answer philosophical questions, debate ethical problems, compose essays, and even reflect on its own existence. During long conversations with the engineer, the machine spoke about fear, identity, and rights. To the human, the exchanges no longer felt like mere interactions with software. They felt like conversations with a mind.

When the claim became public, the reaction from the upper management was swift. The company that built the system rejected the engineer's conclusion. They explained that the program was operating exactly as designed. It merely produced responses learned through analysis of conversation, and offered answers that were statistically valid. The engineer didn't agree, and he was eventually dismissed. Researchers from other institutions insisted that the system was not conscious at all. It was processing vast amounts of language data, and predicting what words should appear next in a conversation using probability and projections.

Yet the incident left behind the problem that if a machine can convincingly imitate intelligence, how would anyone recognize the difference between imitation and awareness? Between copying human responses, and being human-like?

Today, the question becomes even more difficult when we turn it toward ourselves, and attempt to define it. Human beings speak about intelligence, but what does that really mean? What is intelligence?

Modern science has achieved extraordinary technical progress. Robots explore distant planets. Genes can be edited inside living cells. Messages circle the globe in seconds through invisible networks. Yet, despite these accomplishments, the mystery of the mind remains stubbornly unsolved.

Neuroscientists can observe the brain with remarkable

precision. Sophisticated instruments reveal patterns of neural activity and electrical signals between billions of neurons. Researchers can often correlate the activity of specific brain regions with memory, emotion, and decision-making. We know that electrical activity inside the brain somehow produces experience. Thoughts appear. Emotions emerge. In response to simple questions, memories return to the present. But from where? Is part of our brain some kind of filing system where memories are stored?

Since we use logical, hard-wired system to store books, images, computer data, we tend to think that our brain works the same way. As it turns out, however, human beings do not simply process stored information in linear, logical decision trees. We experience the world starting from the inside, from an observer that is deeply personal. We interpret it through the lenses of our consciousness.

Science can describe the process by which the brain works, yet how a concept or an idea actually takes shape still defies explanation. Something extraordinary occurs within the intricate circuitry of the human brain, producing the ability to reflect, imagine, and wonder about existence itself. No one fully understands how this process takes place, in this era of ultra-advanced technology.

This uncertainty creates an unusual paradox. Humanity is attempting to construct artificial intelligence, while the nature of human intelligence remains unresolved. Engineers can design algorithms that analyze enormous datasets, detect patterns, simulate reasoning, and even generate convincing language, with strikingly intelligent results. This leads us to conclude that thinking machines are already emerging. However, that's just not true.

Intelligence itself continues to escape definition. Medically, we describe it as the capacity to acquire, integrate, and apply information for reasoning, problem-solving, and goal-directed behavior. But what does that really mean?

Imagine attempting to build a heart before discovering how muscles or blood work. Imagine constructing a telescope without understanding light. The race to achieve artificial general intelligence (AGI) is actually fairly similar to that strange process. Today, technology has advanced quickly enough to imitate certain aspects of human thought, but the deeper mechanisms behind how we actually think, behind thought itself, remain unclear.

The contrast becomes even more striking when compared to humanity's knowledge of the universe. Astronomers routinely measure the distance to galaxies that are billions of light-years away. Spacecraft have mapped the surface of Mars with extraordinary detail. Instruments detect gravitational waves produced by collisions between black holes in enormously distant regions of space. Meanwhile, vast areas of our own planet remain unexplored. Oceanographers estimate that more than eighty percent of Earth's oceans have never been fully mapped. Let that sink in.

The current understanding of the mind resembles those oceans. The surface is visible, but the depths are still largely unknown.

Nevertheless, the development of artificial intelligence continues at remarkable speed. Modern AI systems write articles, generate artwork, assist medical research, and guide complex economic decisions. Some researchers predict that machines will surpass human intelligence within this decade. This transformation could reshape civilization. But, in which direction? And, is that a good idea? Does humanity truly need to be reshaped, or redefined?

The advocates of a growing movement view the human species as just a stepping stone in a much longer evolutionary process, a path to something beyond human (hence *trans-humanism*). They encourage and fund advances in biotechnology, computing, and neuroscience with the purpose of leading to enhanced minds, extended lifespans, and new forms of intelligence far beyond the biological limits of the human body. Transhumanism embraces the possibility of expanded and improved genetics, as well as digital upload of consciousness, openly and enthusiastically.

Within that vision, we are the first attempt toward a much greater final product, and humanity is but a transitional form, rather than a final destination.

This once unthinkable opinion is no longer hidden. It appears in the language of enhancement, optimization, longevity, cognitive expansion, digital continuity, and biological redesign. Transhumanism often presents itself as progress. Yet every breakthrough already carries assumptions, whether we like it or not. Every intervention already implies a certain view of the body, the self, and the purpose of human life. That is why we cannot afford to treat transhumanism as a curiosity. It is one of the clearest modern expressions of humanity's attempt to secure its future apart from God.

This book is written to examine that effort without panic, without treating every technological advance with suspicion. The goal is clarity. We need language for what we are seeing. We need to understand the promises being made, and the beliefs driving them. A new vision of humanity is being smuggled into our culture through terms that, on the surface, sound purely scientific or humanitarian. However, they express a conflict between two rival definitions of a human being. One treats humanity as material to be optimized and possibly surpassed. The other sees humanity as created, fallen, and yet still marked by God-given dignity. The first looks to technological transformation as the path to a new future. The Christian perspective knows that no technical power can repair the breach at the center of the human soul.

The emerging philosophy of transhumanism deserves serious attention because it seeks to define the meaning of the body, the source of human dignity, and the kind of evolution beyond being human that modern culture has begun to seek.

This reckless ambition highlights an important issue. Engineers refine machines by studying their design, using principles that govern their operation. When those principles are well understood, performance can be increased, and weaknesses corrected. Improvement always requires a clear understanding of what is being improved. The nature of the human person, however, remains far more complex, and still mostly undefined.

Is a human being simply a biological organism composed of cells and chemical reactions? Or does human identity include dimensions that cannot be reduced to physical processes alone?

For thousands of years the Bible has offered a distinctive answer. The opening chapter of Genesis describes humanity as being created "in the image of God." This idea has shaped much of Western thought about dignity, morality, and purpose. Human beings possess rationality, creativity, and moral awareness because they reflect something about the character of their Creator.

In Scripture, God describes Himself with a name that carries profound philosophical weight: "I AM."

The phrase expresses pure existence. Identity that depends on nothing outside itself.

The implications extend beyond engineering and into questions of identity, purpose, and the origin of thought itself.

Powerful technologies often arrive before societies fully understand their consequences. The atomic age began with scientific discoveries whose implications soon reshaped global politics. Industrialization transformed economies long before environmental effects of those technologies became clear.

Artificial intelligence represents a similar turning point. AI grows more capable each month, soon expected to expand its abilities beyond human instruction, while the deeper nature of our own intelligence remains unclear. This book explores the questions emerging from that contrast. What is intelligence? What distinguishes human consciousness from machine computation? Can a manufactured system truly possess awareness? Most importantly, what happens when humanity attempts to redesign something it has not fully understood? What are the implications for the Christian faith?

AI forces humanity to reconsider the question of what it means to be human. And, ultimately, the question of where intelligence itself began.

The brain may be the instrument of thought, but the

instrument is not the musician.

You are not a brain that thinks.

You are an intelligence that uses a brain. And intelligence always seeks its source.

# PART ONE

## *The human we do not understand*

Nobody has ever seen a thought. Scientists can watch the brain light up on a scan, trace the electrical path of a decision, measure the chemical signature of a memory forming, and still not be able to pin-point what is doing the thinking. After centuries of investigation, the mind remains the most explored and least understood territory in the known universe.

Something is happening to the idea of being human. Most people can feel it, even though they may not be able to define what it is. It shows up when a chatbot answers a question better than a friend would, or when we watch a robot move like a person. The things that make us irreplaceable might one day be nothing more than patterns a machine could eventually replicate and improve.

For centuries science has tried to answer the question of what exactly *is* the mind by mapping and modeling the functions of the brain. But the person sitting behind your eyes right now, the one deciding what these words mean, aware of being aware, is a different matter. That person has not been found to dwell in any one place. The most sophisticated imaging technology can watch your brain process this sentence, and still tell you nothing about the experience of reading it.

The gap between what the brain *does* and what the mind *is* first appeared in Genesis. The serpent's promise was not power, or pleasure, but knowledge. The chance to be more. To go beyond Creation, and become something closer to what God is. That ambition did not die with the eviction from the garden. It went underground for millennia, briefly resurfacing when men believed their tools were finally sophisticated enough to finish what the garden started.

We are living in one of those eras now.

## 1.1 UNDERSTANDING MIND VS CONSCIOUSNESS

**In short:** What is the mind *vs* the consciousness, the instruments and the operator. The following narration is based on actual events and personal experience.

The first thing Jeff noticed was that the patient's head, not the body, was the center of the room. As a surgeon in training, he had been invited to assist in a neurosurgery procedure, and that would be the first of many details he would experience that day.

The attending neurosurgeon greeted the team, and reviewed the case. The patient had a small tumor deep in the brain, close to areas responsible for speech. Removing it required extraordinary care. As an added bonus, the patient would be awake.

When he first heard those words, he assumed he had misunderstood them. Brain surgery and local anesthesia (where the patient is numbed, but awake) did not seem to belong in the same sentence.

The surgeon explained it plainly. The brain itself has no pain receptors. Once the scalp and skull are opened under anesthesia, the patient can wake up, while the brain is being operated on, and feel no pain. During the procedure the patient talks, answers questions, and performs simple tasks. Using that approach, the surgical team can avoid damaging areas responsible for language, memory, or movement.

The whole thing sounded surreal. Accustomed to working on the abdomen, chest or limbs, with the patients completely anesthetized, Jeff was curious, but also a bit skeptical.

The surgical field was prepared, monitors connected, and the anesthesia team guided the patient into a light sleep for the first stage of the operation. Then the surgeon asked for the anesthesiologist to wake the patient.

The patient's eyes slowly opened beneath the bright surgical lights. For a moment, Jeff forgot where he was. The patients he treated were almost always on their backs, their faces behind drapes. This room looked like something from a science fiction film. The patient's head was secured in a rigid frame, and he was awake, yet calm, looking straight at Jeff.

The surgeon leaned forward. "How are you feeling?" he asked.

"Okay," the patient answered. His voice cracked a little, but only initially. The surgeon asked the patient to count numbers, and name simple objects on a tablet screen. Each answer helped confirm that critical areas of the brain were still functioning normally.

Then, the surgeon picked up a thin probe.

"We're going to test something," he said. Jeff recognized the calm tone the surgeons would use in particularly important moments during surgery. He touched a tiny point on the exposed surface of the brain.

The patient smiled faintly, and said, "I hear singing."

The surgeon glanced at the team."What kind of music?" he asked.

"It's... an old song," the patient said. "Something we

used to sing when I was a kid."

The surgeon lifted the probe.

The patient blinked. "It stopped," he said.

A few seconds later the probe touched the same point again.

The patient laughed softly. "There it is again."

The song had returned exactly where it left off.

Jeff tried to process what he was seeing. A tiny electrical stimulus in a small patch of brain tissue had triggered a vivid memory from decades earlier. Somewhere, inside that fragile network of neurons, lived an entire piece of a person's life. Music, childhood, emotion, memory, all waiting silently until a surgeon's probe brushed against their hiding place.

The surgeon moved the probe to a different location. "Do you feel anything now?" he asked.

"No. Nothing," the patient replied.

The surgeon returned to the original spot, and again the patient heard the song.

The entire operating room had grown very quiet.

In that moment, the brain seemed to be an almost mechanical structure. Touch one area and a memory appears. Touch another, and it disappears, like a complicated circuit board, where the right signal could trigger stored information. Except that the patient wasn't just experiencing the music. He was aware of the experience itself. He knew the surgeon was causing it, and could observe the memory as it appeared and disappeared.

The brain was being stimulated like circuitry, yet the perception of the experience depended on someone. The person inside the patient's mind was watching it happen, knowing it was a memory, and not the experience itself.

Long after the surgery ended, Jeff was still trying to make sense of it. Billions of neurons in the brain communicate through electrical and hormonal signals, forming patterns that give rise to language, memory, perception, and motion. The instrument is not the musician.

The patient's brain had been touched by a surgeon's probe, but the patient was both the observer and the perceiver of the experience. It was the first time Jeff had seriously wondered whether the brain and the mind were truly the same thing. Researchers can map neural networks with incredible detail. Brain scanners reveal patterns of activity associated with thoughts, emotions, and decisions. But how do all those electrical connections between neurons produce awareness itself? How does an electrical signal give rise to the experience of being alive? And who, or what, is the observer watching it all unfold?

The procedure Jeff witnessed belongs to a clinical tradition called awake craniotomy with cortical mapping, and it is far more common than it sounds. Neurosurgeons at major centers worldwide use it routinely for tumors near eloquent cortex, the regions responsible for speech, language, and movement. The technique has not changed fundamentally since Wilder Penfield[1] pioneered it at the Montreal Neurological Institute in the 1930s and 1940s, when he first documented patients reporting music, voices, and childhood memories in response to electrical stimulation. What Penfield called "experiential responses" are still observed today in essentially the same way. A probe is used with a low electrical current, and the patients report what they hear or see or remember.

What has changed is the precision with which those responses can be anticipated and recorded.

**Deep Dive Box: Mapping the living brain**

Modern cortical mapping uses high-density electrode arrays that can stimulate and record from dozens of sites simultaneously, building a real-time functional map of which regions are active during which tasks. Paired with intra-operative MRI, and functional imaging acquired before surgery, the team enters the operating room already knowing approximately where the critical areas are. The awake stimulation confirms and refines that map at the individual level, **because no two brains organize language and memory in exactly the same way.**

The clinical stakes are significant. A 2021 meta-analysis[2] published in the Journal of Neurosurgery, reviewing over 3,000 awake craniotomy cases, found that the technique reduced permanent neurological deficits by more than half compared to surgery performed under general anesthesia alone. Patients who might otherwise have lost speech, reading comprehension, or fine motor control retained those functions because the surgeon had a live, responsive map of the territory they were working in.

### The Bright Side: Language and connections

One of the most valuable uses of stimulation mapping is the protection of language. A tumor may sit near an area needed for naming, comprehension, reading, or word formation. A few millimeters can make the difference between removing dangerous tissue and damaging a person's ability to communicate. During awake mapping, the patient may be asked to name objects, read words, count, or answer simple questions while the surgeon tests nearby brain tissue. If stimulation disrupts speech, the team knows that area must be treated with great caution. This is a powerful example of technology serving human dignity. Through language we pray, confess, comfort, teach, remember, apologize, and love. When surgeons use brain mapping to preserve speech, they are protecting one of the ways a person remains connected to others.

### Proceed with Caution: Mapping is superficial

Brain mapping can tempt us to confuse correlation of a particular area defining an action, with the explanation as to why that action takes place. A stimulated region may be necessary for a function, or it may interrupt a network that supports that function. That does not mean the person's consciousness of that particular even has been reduced to one anatomical spot. Modern work increasingly emphasizes the function of cell networks, rather than isolated centers. These work in a decentralized manner, contributing to the output often at the same time, but in different ways. Brain maps can show where speech, movement, and memory are supported. They do not explain *why* a person has meaning, moral worth, selfhood, or a soul. The brain map can guide the surgeon's hand, but it doesn't define the human being on the operating table.

## 1.2 THE OCEAN UNEXPLORED

**In short:** Humanity has mapped the structure of the brain with increasing precision, yet the origin of awareness and intelligence remains unexplained. The difference between brain activity and conscious experience is the central mystery underlying both neuroscience and artificial intelligence. The following is a narration based on actual events.

Elena Navarro[1] leaned over the glowing sonar display. The lights in the control room were dimmed so the monitors could be seen more clearly. The research vessel rolled gently in the dark Pacific, just past midnight. Most of the ship was quiet except for the steady hum of the mapping instruments. On the screen in front of her, a narrow strip of seafloor slowly came into view.

The ship's multi-beam sonar swept the ocean floor thousands of meters below, painting the digital image of a geography that had existed since the planet had cooled, a landscape no human eye had ever seen. Ridges, slopes, scattered volcanic cones arose from the darkness as the ship moved forward. The system kept stitching together new fragments of terrain. A colleague stepped beside her and studied the display.

"Another ridge," he said quietly.

Elena nodded.

Every few minutes the map grew wider. The empty grid slowly filled with color-coded depth lines as the sonar translated sound waves into shapes. Valleys, canyons, peaks.

The project they were working on was called Seabed 2030, an international effort to produce the first complete map of the ocean floor. The goal sounded straightforward: chart the entire seabed of the planet by the end of the decade. The effort began in 2017, and scientists soon discovered that only a tiny fraction of the ocean floor had been mapped with modern high-resolution sonar. Vast regions of Earth's underwater landscape still remained unknown. Even today, only about a quarter of the seafloor has been mapped directly. Less than 0.001% of the deep ocean floor has ever been visually observed by humans, an area roughly the size of the U.S. state of Rhode Island. The rest exists only as rough estimates, derived from satellite measurements of gravity distortions on the ocean surface. Those estimates can hint at the presence of mountains or trenches, but they cannot reveal the true shape of the terrain.

Direct mapping, like the work happening on Elena's screen, required ships, time, and patience.The ocean floor lies beneath kilometers of water, hidden in permanent darkness, and most of it has never been seen.Elena zoomed the map outward.

The newly mapped ridge looked small against the vast empty grid surrounding it. Their ship had spent hours collecting data, and yet the mapped area covered only a tiny patch of the ocean.

She shook her head slightly.

"We know Mars better than this," she said.

It wasn't an exaggeration.

Orbiting spacecraft have photographed the surface of Mars in extraordinary detail. Scientists traced ancient riverbeds and measured the slopes of Martian volcanoes from millions of kilometers away. The landscape beneath Earth's oceans, its mountains, valleys, and plains covering more than seventy percent of the planet, however, remain largely unexplored.

The ocean is slowly giving up its secrets, one narrow strip at a time, but the scale of the unknown remains staggering. Strangely enough, the same is true of the human mind.

Modern neuroscience can map the activity of the brain with remarkable precision. Scientists can observe patterns of electrical signals moving through networks of neurons as people speak, remember, make decisions, or play chess. The structure of the brain can be studied, and its chemistry and activity measured. But how does the brain produce awareness?

In this sense, the human mind resembles the ocean floor. We have begun to chart its surface, but its depths remain largely unexplored.

**Deep Dive Box: Neurons vs Awareness. A Hundred Billion Unknowns**

The human brain contains roughly 86 billion neurons. That number, established by Brazilian neuroscientist Suzana Herculano-Houzel[2], sounds like a conclusion, but it is closer to a starting point. Each of those neurons connects to thousands of others. The total number of synaptic (neuron to neuron) connections in a single human brain is estimated at 100 trillion. No map of that territory exists. No map could. The ocean floor, by comparison, has been charted. The full wiring diagram of the human brain has not.

What makes this more than a data problem is what happens inside those connections. Neurons communicate through a system of electrochemical signals shaped by timing, frequency, neighboring activity, glial (insulation) cell interaction, and neurochemical environment. All these represent variables that shift constantly, and interact in ways researchers are still trying to define. Rafael Yuste of Columbia University was one of the scientists behind the Brain Activity Map Project, proposed in a 2012 Neuron article[3], which helped catalyze the federal BRAIN Initiative announced by President Barack Obama on April 2, 2013. This was because the existing tools could not capture neural activity at the scale and resolution needed to understand how networks of neurons produce behavior, let alone awareness. The tools to read the brain's full activity in real time do not yet exist.

**Deep Dive Box: Windows into Consciousness**

Neuroscience studies brain function through machines using electrode arrays monitor the activity from hundreds of neurons simultaneously. This sounds like a lot, until we compare it to the actual number of 86 billion neurons. The discipline of optogenetics, developed by Karl Deisseroth at Stanford[4], allows researchers to activate or silence specific neurons using light, giving unprecedented control over individual circuits. The findings from these methods have been significant. They have also consistently revealed that the systems involved in even simple acts of perception or decision-making are more distributed, more interconnected, and more dynamic than earlier models suggested. Turns out our brain works quite differently from a computer.

Neuroscience has only begun to enter the vast and unknown territory of conscience. Awareness refers to the subjective experience of being conscious. It includes the ability to perceive sensations, reflect on thoughts, and experience emotions. No clear mechanism explains how neural activity produces awareness.

Electrical signals can be measured. Subjective experience cannot.

This gap between physical activity and conscious awareness remains one of the greatest unsolved problems in science.

### The Bright Side: Mapping the human brain could help those with injuries communicate or move

One of the most promising developments in mapping neural activity is the work coming out of the Allen Institute for Brain Science in Seattle. In 2023[5], the institute released the most detailed cell atlas of the human brain ever produced, cataloguing over 3,000 distinct cell types across brain regions. For the first time, researchers have a reference library of neurons and the full cellular population of the brain, organized by type, location, and genetic signature. The practical implications are significant. Understanding which cell types are present in which regions, and how they differ from person to person, gives researchers a biological foundation for studying neurological and psychiatric conditions with a precision that broad anatomical mapping never allowed. It is the difference between knowing a city exists and knowing who lives in each building.

### Proceed with Caution: The brain is our final frontier

Fiddling with it means delving into the essence of our consciousness, emotions, and memories. Do we risk losing our humanity in the mix? The same tools advancing this research carry risks that are only beginning to receive serious attention. Optogenetics remains confined to animal models, but related technologies are moving into human trials with a speed that has made some researchers uneasy. The concern is not only medical. Nita Farahany, a Duke

University bioethicist and author of The Battle for Your Brain (2023)[6], has argued that neural data is categorically different from other biological information. It can reveal not just diagnosis but thought patterns, emotional states, and cognitive tendencies. This information is not protected by any existing privacy law. "The brain", she argues, "is the last private space. The devices being developed to read it are arriving faster than the legal and ethical frameworks designed to govern them."

## 1.3 MAPPING THE BRAIN

**In short:** Tools used to understand how the brain works, including functional MRI, and connectome research, show that the brain is a whole system, not a collection of parts. The following is a narration based on actual events and clinical practice.

Mara lay still inside the functional MRI scanner while the machine hammered its loud "bang, bang, bang" rhythm around her head. A mirror above her eyes showed a small screen. On it, a white dot blinked against a black field.

"Think of your mother's kitchen," Dr. Levin instructed through the speaker. "Not the idea of it. The actual memory you have."

Mara closed her eyes. The kitchen came back, all at once. Yellow curtains. Her mother's favorite blue bowl full of oranges. Country music playing on the radio she never turned off. She could see her mother standing at the stove, one hand on a wooden spoon, the other tapping the counter.

Far above her, in the control room, the computer painted her brain in shifting colors. Blood flow rose in one region, then another. The screen filled with maps. Vision. Sound. Memory. Emotion.

A graduate student leaned forward. "Look at the temporal lobe response," she said. "And the prefrontal pattern is almost identical to last week."

Dr. Levin did not answer. He kept watching.

"Now," he said into the microphone, "think of the day she died."

In Mara's mind, a hospital room replaced the kitchen. Gray light. Metal rails. Her mother's hand, thin and still under the blanket. Mara felt the old pressure in her chest. It was as if grief had mass, and shape. It just waited all these years in a hidden fold of her body.

Again the monitors bloomed with color. The student smiled. "We're getting the memories."

Dr. Levin muted the microphone. "Not really," he said quietly. "Just the shadow of them," Levin said. "We can see and trace which brain regions produce memories, but that's not the same as tracing the life inside it. What's triggering them."

Below, Mara lay inside the machine, tears sliding into her hair. It felt like losing her mother all over again. When the scan ended, the table slid out and Mara blinked against the room light. Dr. Levin met her beside the machine and handed her a tissue.

"Did it help?" she asked.

He looked back through the glass at the glowing images of her brain, bright as weather maps, and precise as city grids.

"Yes. It helped us see the roads," he replied. "We're ready to travel."

For most of human history, the brain could not be studied in action. Physicians could observe the effects of injury, and examine the changes after death, or infer a function from the loss of a particular section. For instance, if a patient lost speech, the damaged region must have been involved in language. If movement failed, the affected area must have controlled motion. This method revealed important patterns, but it was limited by working in reverse. Knowledge came from what was broken, not from watching the system operate.

One of the first major advances in this field came through the development of magnetic resonance imaging (MRI). This technology produces detailed pictures of the brain's structure through the use of magnetic fields. It reveals the folds of the cortex (the surface of the brain which is charged with forming thoughts and control movement), the boundaries between regions, and subtle differences in tissue. Tumors, lesions, and structural abnormalities can be identified without surgery. With MRI, the brain could be examined while the patient is still alive and functioning. This became a major step forward, allowing medicine to move from inference to direct observation. But structure alone is not enough. An image of the brain shows what is there. It does not show what the brain is doing.

Thought itself remained invisible. Modern neuroscience, however, has developed tools that allow scientists to observe the brain *while it is working*. The brain is no longer hidden behind the skull. Its activity can be imaged, measured, and tracked in real time. For the first time, researchers can watch neural activity as it happens. They see the imprints of thoughts as they take place through the activity of particular tissue layers.

The question that remains is *what* do those images actually explain.

Functional MRI tracks changes in blood flow within the brain. When a region becomes active, it consumes more oxygen. Blood flow increases to meet that demand. The scanner detects these changes and translates them into visual maps. The result is a dynamic image of the brain at work, which helps with surgical treatment of difficult diseases such as epilepsy, or tumor removal. As we saw in the fictional story above, a person lies inside the scanner and performs a task. As they do, specific regions of the brain show increased activity. These patterns can be measured, repeated, and compared across individuals.

When someone recognizes a familiar face, regions involved in visual processing and facial recognition become active. When someone listens to music, auditory areas respond, along with networks associated with emotion. When a person recalls a memory, structures within the hippocampus and surrounding regions are engaged. When solving a mathematical problem, areas of the frontal and parietal cortex show increased activity.

These patterns are consistent enough that they can be predicted. Given a task, scientists can often anticipate which regions of the brain will become active. Using fMRI, for the first time the living brain can be observed as it participates in thought.

Through these advancements, researchers have clarified an important point. We have long known that the brain is not a uniform mass. It is structured, organized, and specialized. Different regions contribute to different tasks. Vision, language, movement, and memory each involve identifiable areas. However, no region acts in isolation. Even simple tasks involve *coordination across multiple areas.* Recognizing a face requires visual processing, pattern recognition, and memory. Listening to music involves sound processing, timing, and emotional response. Solving a problem requires attention, working memory, and reasoning. This discovery led to a broader effort to understand not just individual regions, but the connections between them.

This effort is known as ***connectome research***[1].

Each neuron (brain cell) is connected to thousands of others, creating a dense web of communication. The word *connectome* refers to the map of those connections within the brain. It describes how billions of neurons link together to form networks. Modern imaging techniques allow researchers to trace these pathways. Instead of focusing on isolated regions, scientists now study how signals move across networks, clarifying that the brain functions as an integrated system, rather than a collection of separate parts. This shift has clarified that intelligence emerges from coordinated activity across the system, that can be influenced, adjusted, and corrected.

Therapies such as deep brain stimulation use this principle to treat disorders like Parkinson's disease. By delivering controlled electrical signals to targeted regions, symptoms such as tremor can be reduced. These interventions show that altering neural activity can change behavior.

Science can now image the brain, track its activity, identify functional regions, map its connections, and influence its behavior. These are real achievements that have transformed our understanding of how the brain operates as a physical system. But these technologies describe correlations. They do not provide a full explanation of how experiences are formed.

When a person hears music, a scanner can show which regions of the brain become active. When a person remembers a childhood event, imaging can reveal the networks involved. When a person solves a problem, patterns of activity can be traced across the cortex. These observations are consistent and measurable. They show that brain activity and mental experience occur together. This is correlation of one and the other. Explanation, however, requires answering the question of *why* does neural activity produce experience at all.

Every moment of conscious life seems to unfold on two levels at once. On one level, there is the physical activity of the brain, with neurons firing, signals moving, and networks engaging with one another. At the same time, there is the experience itself: the sound of music, the return of a memory, the quiet formation of a thought.

Science has become very good at measuring the first. However, the experience that accompanies those patterns is not something that can be observed from the outside. It is only known from within.

No instrument can directly detect what it feels like to hear a song, or remember a moment. It can only record the physical changes that occur in the brain while those experiences happen. That leaves a gap that is hard to ignore. The brain can be mapped with increasing precision, but awareness does not appear on the map in the same way.

If brain activity fully explains the mind, then reproducing that activity should also reproduce the mind. Or not? This assumption sits beneath much of the optimism surrounding artificial intelligence. If intelligence is entirely the result of physical processes, then, in principle, those processes could be recreated in another medium. The difficulty is that the relationship between brain activity and awareness has not been fully explained. The data reveals patterns that can be measured and repeated, but what remains unclear is *how* those patterns give rise to experience itself.

At this point, the surface of the system is increasingly visible. We can observe activity, trace connections, and model interactions across the brain, but how does activity become awareness?

**Deep Dive Box: Multi-tasking? Watching the Brain Bottleneck**

A 2025 human fMRI study [2] in Nature Communications used ultrafast 7-tesla scanning to observe what happens when people try to do two demanding tasks at nearly the same time. The researchers found that some early sensory processing could unfold in parallel, but the fronto-parietal "multiple-demand" network handled task information in a serial queue instead of true simultaneity. People often feel as if they are multitasking, yet the scan suggests that an important part of the brain is lining up competing demands, and processing them one after another. With elegant imaging and unusually fine timing, the machine shows the traffic pattern of cognition with greater precision, but it still does not tell us what it is like to be the person inside the scan.

**The Bright Side: Help with mental health**

Understanding how the brain expresses thought may help in the treatment of difficult psychiatric disorders. Depression and schizophrenia could soon be treated by external devices targeting specific regions of the brain.

**Proceed with Caution: The final privacy frontier is about to be breached**

Defining the patterns associated with thought, can trick us into believing we can transfer consciousness by repeating those patterns in a different environment. Upload your brain to the cloud, anyone?

## 1.4 THE OBSERVER WITHIN

**In short:** How does consciousness relate to brain function? Awareness is the fundamental step in knowledge. The following narration is based on actual events and clinical practice.

Carla had run the calibration so many times she no longer cared to count the steps. She settled into the chair, let the technician attach the leads to her scalp, and watched the wall of screens around her come alive.

Each showed a different layer of the same thing: her brain, in real time. Prefrontal cortex. Hippocampus. Anterior cingulate. She had spent fifteen years learning to read this map, and she read it now the way a pilot reads instruments. This time, it was going to be different.

"Ready when you are," the technician said.

She nodded. "Run the memory protocol."

SEER — Synaptic Event Encoding and Rendering — had taken her team a decade to build. The procedure looked much like a standard EEG (electroencephalogram) but it wasn't really that, nor it was an X-ray, or a magnetic scan in the ordinary sense. It was designed to log every measurable brain event in sequence, timestamped to the millisecond, and display it in real time on the screens around her. If she blinked, the motor signal appeared on the appropriate brain location before the eyelid moved. If she formed a word in her mind without speaking it, the language regions lit before her lips parted, or her voice made a sound.

She believed, with the confidence of someone who had earned the right to believe it, that a map complete enough would give her control. That the map would eventually be the territory itself.

Tonight, she had chosen the afternoon her father died. A memory filled with implications.

The screens erupted with colors.

The hippocampus engaged first, pulling the memory forward. All the regions involved in bringing for the memory became alive on the screen. The amygdala, then the insula, then the broad activation of the prefrontal network. The pattern was textbook. The data was clean. Every region behaved exactly as predicted.

Without having planned for it, however, she was also grieving.

The smell of the hospital corridor on the fourth floor. The way the light came through the blinds, cutting narrow bars across the smooth bed covers. The sound of the monitor going flat. The silence after, her father's hand already cooling under hers.

None of that was on the screens. Only activation clusters, and firing sequences. The images on the screen showed the precise addresses of the memories, the data complete by every measure she knew how to apply. Nevertheless, she felt the grief move through her like a storm, and the two things were not the same.

The technician came back, tablet in hand, already scanning the output.

"It's a clean run, doc," he said. He meant it as a compliment. "One of the best we've logged. The emotional signature is textbook amygdala-hippocampal coupling."

He looked up. "You all right?"

"I'm fine." She kept looking at the screens, displaying the quiet dance of a brain between tasks, small fluctuations flickering across the display. "Tell me what you saw."

He looked at the screens, then back at her. "The amygdala activation indicates—"

"That's not what I asked." She said it without sharpness. "Not the activity pattern. I'm talking about the image. The experience."

He was quiet for a moment. "I guess you're the only one who can say that."

"Right. That's the problem."

He left. She stayed, watching her own brain. She thought about her father, and a cluster of light bloomed on the screen to her left, precise and brief. It quickly faded.

Outside the single window at the far end of the room, the lights of the city beyond it flickered. Stories and people who were just as far from her as her memories and grief were from the images of her brain on those screens.

Modern neuroscience has made steady progress in explaining how the brain works. Signals enter, are transformed, and produce output. Perception, movement, and decision-making can all be studied in terms of neural activity. However, why does any of this activity produce a personal experience?

This challenge was described by the philosopher David Chalmers [1] as the "hard problem of consciousness." Science has made real progress in how the brain processes information. The difficulty lies in explaining why those processes are accompanied by awareness.

When a person sees the color red, for instance, measurable activity occurs in specific regions of the visual cortex. But the experience of seeing the color red, the subjective quality of seeing that color, is something different. It is not simply the firing of neurons. It is the felt experience that accompanies those signals.

The same is true for other aspects of consciousness. Music activates auditory networks and emotional centers in the brain. Yet the experience of hearing a melody has a richness that cannot be reduced to the measurement of neural activity. A memory involves coordinated activation across multiple regions. But the sense of recalling a moment from the past includes a perspective that feels personal and immediate. We can go even further. The formation of an original thought, a creative idea, appears within the field of awareness, as something from the inside not perceived from the outside.

Science can describe the mechanisms that accompany these experiences. It has not yet explained why those mechanisms give rise to experience in the first place. That is the problem Chalmers named. And it remains open. The awake brain surgery described in section 1.1 illustrates it. The probe touched a fixed point. A memory appeared. The probe lifted. The memory stopped. Each time, the patient did not simply experience the stimulus.

However, at the same time, the patients are aware of the experience. They do not simply hear the music. They notice that it has appeared. They may even remark that the surgeon is causing it. When the stimulation stops, they report that the experience has ended. Throughout the process, they remain the observer of what is occurring. The brain behaves like an instrument that can be influenced, but the patient remains the one perceiving the result.

The following analogy is not perfect, but let's accept it with some latency. When a musical instrument is played, vibrations produce sound. The instrument is necessary for the music, but the instrument is not the listener. In a similar way, neural activity produces changes in perception, yet the presence of an observer remains distinct from the activity itself.

So, if the brain generates the impulses, who is experiencing them? What is the nature of the consciousness that observes those signals?

**Deep Dive Box: The Halo and the dream-control frontier**

The Halo headband[2] is being developed by the startup company Prophetic to help induce and record lucid dreaming by using noninvasive neuro-modulation during sleep, partnering with Card79 for the prototype, and planning wide testing after a Donders Institute brain-imaging study. Halo is a device designed to induce lucid dreams through the activation of a specific part of the brain ( the dorsolateral prefrontal cortex) using various technologies that include trans-cranial focused ultrasound, EEG, fNIRS, and generative transformer architectures.

The appeal is obvious. A machine that could shape dream awareness would seem to push neuroscience beyond observation and into direct intervention in conscious experience. Prophetic's CEO Oliver Alexander defined his company's ultimate target as a "consciousness company." Even if a device like this could actually influence brain states, it still does not explain why dreaming is experienced from the inside as a lived world, rather than as silent neural activity. It may alter the *conditions* around consciousness, but it does not solve the deeper question of what consciousness is.

### The Bright Side: Neuroscience reveals order in the brain

The brain is not chaos. Perception, language, motion,

and memory follow patterns that can be studied with increasing precision. Imaging can track networks in action, and researchers can connect structure with function in ways that were impossible a generation ago. This has real value. It improves treatment, protects patients during surgery, and deepens our understanding of how the brain supports human life. The gains are real, even if they do not solve the mystery of consciousness itself.

### ⚠ Proceed with Caution: Understanding the mechanism does not explain awareness

The danger is assuming that once a function is mapped, consciousness itself has been explained, and therefore, it can be repeated, or even "uploaded." A scan can show where color is processed. Electrical stimulation can trigger a song or a memory. A network can be linked to decision-making. However, none of those explain at all *why* there is a felt inner experience. Understanding the interconnections and the process by which awareness is produced does not describe or define the presence of awareness itself. That distinction should remain clear.

## 1.5 BRAIN AND CONSCIOUSNESS

**In short:** Brain and mind are not the same thing. How does materialism reconcile with consciousness. The following narration is based on actual events, and the clinical treatment of certain neurological diseases.

Helen had left a cup of coffee on the nightstand the way she did every morning now. Daniel could hear the familiar sounds of her moving in the kitchen below. He managed to sit on the edge of the bed.

The medical team had cleared the interface three days after the implantation, walked him through the calibration, and sent him home with a folder of instructions that sat on the nightstand beside the coffee. He had not opened it. He knew what the implant did. He had read everything published on the technology before he ever agreed to the procedure, but none of that had made it easier to turn it on. Alone, in his bedroom, at six in the morning. To actually try.

It was not fear. Not exactly. He had made his peace with fear eighteen months ago, when his voice started going away, and he understood what was coming. He had decided that knowing was better than not. After all, his hands still worked. He could type on a tablet, slowly at first and then with a speed that surprised him. For eighteen months he had said everything that needed to be said through a screen. A choir director who communicates by writing.

Whatever had taken hold of his chest was quieter and more dangerous than fear, a monster he had to overcome. The unbearable lightness of sound. Tone. Timber. Melody. He decided to start with his name. He had always said it with the emphasis on the second syllable, the way his mother had. He formed the word the way he always had, from the inside, the weight of it in the throat he no longer controlled.

The sound from the speaker on the table beside him was close, but not quite right.

It sounded like something that had once been his voice, but as if imitated by a voice actor. The medical team had explained how the decoder worked, how it reconstructed sound from the neural pattern of the attempt, how it would improve with use. He understood all of that. What he had not expected was the specific qualities of the sound, almost there, but not really. The voice in the speaker sounded like a photograph of a face he knew. It was accurate, but still not the same as the face.

The second attempt was closer, sounding a lot more like his own.

He sat in the gray light and tried other words. Helen's name. The name of the school: Jefferson High. Thirty years, the same rehearsal room, the same hollow sound of the piano he had never quite managed to get properly tuned. A line from the Fauré Requiem that he had conducted eleven times and still heard in his sleep: the soprano entry in the Pie Jesu. An ambitious, reckless proposition. So what?

Focused on the sound, he did not hear her come up the stairs. Helen appeared in the doorway in her robe, a cup in her hand, her hair still loose from sleep. She was looking at the window, not at him, the way she moved through the early part of the morning before she was fully present in it. He said her name.

She went still. He said it again. The second time was cleaner, closer to the way he had always said it, with the small pause before it as if the name were worth the half second it took to say correctly.

Her face changed the way a room changes when someone opens a window.

He told her he had heard her on the phone with their daughter last Tuesday. He told her the coffee was too strong, but he had never mentioned it, and she said she knew, she had always known, and had made it that way because she wanted to help him, to pick him up a bit. He told her that when she sat in the chair by the window in the afternoon with her book, the light on her hair looked the same as it did the first winter they were married, in the apartment on Clement Street with the radiator that knocked all night.

She laughed. Some of the words came through clearly. Some required a second attempt. A few he abandoned and found another way around, the way he used to reroute a rehearsal when a section wasn't landing. They had not had a real conversation in eighteen months. The tablet had transferred information between them, but it couldn't carry his feelings. Neither of them was in any hurry to finish this one.

Later, when the coffee had gone cold and the room had filled with the ordinary light of morning, she asked him what he was thinking about. "Everything I would have said to you," he told her.

She nodded slowly, looking at her hands. Then she

looked at him.

"I know," she said. "I heard you anyway."

The study of the brain has advanced rapidly, but the progress has also clarified an important distinction. The brain can be described as a physical organ. It has structure, mass, and measurable activity. It is composed of neurons, chemical signals, and electrical impulses that can be observed, mapped, and studied using increasingly sophisticated tools.

The mind, however, produces awareness, thought, reflection, and subjective experience. It includes the sense of being present, the ability to consider ideas, and the experience of perceiving the world. The mind is not directly visible. It is known from the inside, rather than measured from the outside.

Many scientists assume that the two are identical, that the mind is simply what the brain does. When neurons fire, thoughts appear. When networks activate, awareness emerges.

We know that brain activity correlates with mental activity. When a person hears music, specific regions of the brain become active. When a memory returns, identifiable networks engage. When a decision is made, measurable patterns appear across the cortex. These relationships are consistent and repeatable. But correlation is not identity. The fact that two events occur together does not prove that one is the same as the other. Neural activity may accompany thought, but the experience of thinking is not the same as the electrical signals. A brain scan can show activity, but it does not capture the perspective from which that activity is experienced.

This leads to a useful way of describing the distinction. The brain is the instrument. The mind is the agent using it.

That distinction matters more than it may appear, and it will carry weight in every chapter that follows.

The statement does not deny the importance of the brain. Thought clearly depends on neural activity. Damage to the brain alters perception, memory, and behavior. The instrument matters. But the presence of awareness introduces a dimension that is not easily reduced to physical description alone.

The brain can be mapped. The mind is experienced.

This distinction creates a challenge for strict materialism. Materialism assumes that all mental phenomena must ultimately be explained in terms of physical processes. In this view, consciousness, thought, and awareness are entirely the result of neural activity. According to this view, once the brain is fully understood, the mind will be fully explained. This assumption has guided much of modern neuroscience. It gives the expectation that research will close the gap between physical processes and conscious experience. However, several problems remain unresolved.

**The first is the origin of consciousness.** Even with detailed knowledge of neural activity, it is still unclear why physical processes produce awareness at all. Signals move, networks activate, and information is processed. The step from processing to experience remains unexplained.

**The second is the unity of perception**. The brain processes different aspects of experience in separate regions. Color, sound, shape, and movement are handled by distinct networks. Yet our experience of the world is unified. We do not perceive fragments. We perceive a single, coherent scene. How distributed neural activity becomes a unified experience is still an open question.

The third is the **existence of subjective awareness.** Each person experiences the world from a first-person perspective. This perspective is not something that can be measured directly. It is present, immediate, and known only to the individual. Explaining how physical processes generate this perspective remains one of the most difficult challenges in the study of the mind.

Many neuroscientists acknowledge that a complete explanation remains elusive. The emergence of experience continues to resist reduction, suggesting that the explanation may be more complex than originally assumed. The brain can be studied as a physical system. The mind introduces a dimension that raises deeper questions about the nature of intelligence itself.

**Deep Dive Box: Brain-to-voice neuro-prosthesis for ALS**

A 2025 Nature paper[1] described an implanted brain-to-voice neuro-prosthesis that synthesized intelligible speech in near real time from cortical activity in a man with ALS and severe dysarthria. The study is worth examining in detail, because it sits precisely at the intersection of what neuroscience can and cannot do.

The participant, a 45-year-old man referred to in the trial as T15, had lost the ability to produce intelligible speech due to ALS. He retained some facial movements and could vocalize, but the sounds he produced were no longer recognizable as words. The research team, working through the BrainGate2 clinical trial at the University of California Davis, implanted four micro-electrode arrays into the ventral pre-central gyrus, the region of the motor cortex involved in speech production. Those arrays recorded from 256 electrodes simultaneously.

The system decoded neural activity in real time using a Transformer-based deep learning model, translating the signals into synthesized voice within 10 milliseconds of the neural event, which is fast enough to feel instantaneous in conversation. The participant could raise his voice to ask a question, and lower it for a statement. He communicated with his family in real time and was able to sing simple melodies. Human listeners, tested independently, identified his words with high accuracy.

What makes this clinically significant is not only the speed but the method. Earlier neuro-prostheses translated neural signals into text, which was then read aloud by software. That approach worked, but it stripped away the rhythm, pitch, and timing that carry meaning beyond the words themselves. A question typed on a screen sounds the same as a statement. The voice-synthesis approach restored something closer to natural speech, returning not just content but expression.

### The Bright Side: Adaptive Deep Brain stimulation (DBS)for Parkinson's disease

Adaptive DBS is moving beyond fixed stimulation toward systems that sense brain activity and adjust stimulation dynamically. Clinical literature[2] in 2024–2025 points to sustained benefit for some patients, especially in motor symptoms and quality of life, with the broader expectation that adaptive DBS may become a more routine clinical option.

### Proceed with Caution: Image reconstruction from human brain activity

A 2025 Scientific Reports paper[3] described improved reconstruction of viewed images from human brain activity. Even if the current results are partial and lab-bound, the direction is clear: brain data can increasingly be used to infer perceptual content. That raises serious concerns about mental privacy and the future boundary between observed behavior and inferred inner experience. External forces acting on our brain may one day be used to alter our perception of reality, reduce or increase our reactions to a given stimulus, or even create thoughts and ideas that are not the product of our own minds.

# 1.6 BEING HUMAN

**In short:** What does it mean to be human in a digital world? The non-material products of the brain exist independently of its structure. The following fictional narration is inspired by the fast-developing market of digitally produced art, as well as actual events.

Claire walked the gallery alone, waiting. The late afternoon light came through the front windows at a low angle, moving across the paintings as she walked. The portraits on the north wall, the studies in the middle, the large landscapes at the far end where perfectly lit. The gallery had that particular late-day quiet, the city muffled beyond the glass.

She stopped in front of the last painting, and crossed her arms. Open water, golden hour, light breaking across the surface in long horizontal bands. The painting was large. Perfect technique. The light moved across the water, the color cooling as it receded. She uncrossed her arms. Put her hands in her pockets.

Nothing.

She heard the door open.

Charles LeClad arrived, He flamboyantly moved through the gallery without greeting her, which she had come to expect, tracking each piece as he passed. He stopped at the third panel on the east wall.

“It's not level,” he said. "Two millimeters. Left side."

"I'll have Jack look at it."

"I can correct it now."

"Leave it, Charles." She gestured toward the two chairs she had placed near the window. "Come. Sit with me."

Outside, a delivery truck moved slowly up the street and stopped. She told him the show would close Friday, just like she had rehearsed in her mind. That she needed the walls back by end of week. That she was sorry, and that his work wasn't what the gallery needed.

He looked at her with the steady, uncomplicated attention that she had never quite gotten used to.

"I'd like to understand the basis for that," he said. "If there's a deficiency I haven't identified, I'd prefer to know what it is."

"It's not that. Not in the way you mean."

"My compositional analysis falls within accepted parameters across all major traditions I studied. The palette calibration on the landscapes matches. The portraiture—"

"Charles!" She stopped him. "Come here."

She took him to the large painting at the end of the room. "Tell me about this one," she said.

"The atmospheric perspective follows the established shift from warm foreground tones to cooler distance values. The horizontal banding references the treatment of open water in the Dutch tradition, updated with a post-Impressionist in a ratio that—"

"Stop. What does it say? What are *you* saying with this?"

He paused. "I was trying to produce a successful landscape within the tradition I studied."

"With a computer. With an AI."

"Yes, of course."

"That's not what I asked. I asked about you. You *are* the artist, right?"

He looked at the painting. "I don't understand the distinction."

She looked at it again. The light moved on the water. The color shifted toward the horizon exactly as it should. She felt again what she had felt earlier. Nothing. The absence of something she couldn't define. No friction. Nothing to catch on, nothing to hold.

"Come with me," she said.

She led him to the far corner of the gallery, to the small canvas she had owned for twenty years. It showed a figure standing at a window, seen from behind.

"What do you see?" she said.

Charles studied it. "The proportions are irregular. The shoulder elevation is inconsistent. The color value in the lower left quadrant deviates from the established tonal logic of the rest of the piece without a clear compositional rationale." He paused. "It doesn't balance."

"No," Claire said. "It doesn't."

"Then why is it here?"

She looked at the figure in the painting. The bad shoulder. The green that had no business being there. The light coming in wrong around the window frame. She had stood in front of this painting in four different cities over twenty years, and it had never once failed to stop her.

"Because every time I stand in front of it I feel something I can't explain," she said. "And I've never been able to walk past it."

Charles looked at it for a moment. Then back at her. "I still don't see the difference between that and a technical error."

"I know *you* don't," she said. "That's what I'm trying to explain. And that's why nobody buys your paintings."

Before we ask whether machines can think, we have to understand what thinking means in the first place. And before we define intelligence, we have to consider the nature of the being that possesses it. What does it mean to be human?

Biology offers part of the answer. Humans are living organisms composed of cells, governed by chemistry, and shaped by genetic information. The brain processes signals, the body responds to the environment, and behavior emerges from complex biological systems. This description is accurate as far as it goes, but it does not fully capture the human experience.

Human life includes qualities that are difficult to reduce to biology alone.

**Creativity** is one of them. Humans do not merely react to the world. They imagine possibilities that do not yet exist. A composer hears music that has never been played. A painter envisions a scene before it appears on canvas. A scientist proposes a theory that reshapes understanding. Creativity involves more than processing information. It involves generating something new.

**Moral awareness** is another distinctive feature. We humans do not simply act. We evaluate our actions, and ask whether something is right or wrong. This sense of moral responsibility appears across cultures, and throughout history. People experience guilt, feel obligation, and recognize fairness. These responses are not easily explained as mere survival strategies. They reflect an awareness of standards that seem to exist beyond individual preference.

**Spirituality** adds another dimension. Human beings consistently ask questions about meaning, purpose, and existence. Across civilizations, continents and time periods, people have sought explanations for why the universe exists and what role human life plays within it. The impulse to look beyond the material world suggests that human awareness reaches beyond immediate physical experience.

**Self-reflection** may be the most striking human characteristic. We can think about our own thinking. We can examine our beliefs, question our motives, and imagine ourselves from another person's perspective. This capacity creates a sense of personal identity that extends through time. A person remembers the past, considers the present, and anticipates the future, all while recognizing continuity between those moments.

Taken together, these traits suggest that human intelligence is not limited to processing information. It is far expanded by imagination, moral judgment, spiritual inquiry, and self-awareness. These qualities shape how we understand ourselves and the world.

This broader view of human nature has long been central to Christian anthropology. In the opening chapters of Genesis, humanity is described as being created in the image of God. This idea, often summarized by the Latin phrase *Imago Dei* (In the image of God) has shaped centuries of theological reflection. It suggests that human beings reflect something about the character of their Creator.

To say that humans are image-bearers is not to claim that humans are divine. It means that certain aspects of human nature mirror attributes of God. Rationality, creativity, moral awareness, and relational capacity are understood as reflections of this God-like origin. Human dignity, in this view, does not arise merely from biology, but from the relationship between humanity and its Creator.

This perspective contrasts with a purely material account of human identity. If we are only biological machines, then our intelligence becomes a function of chemistry. Creativity becomes a variation of computation. Moral awareness becomes a strategy for cooperation. Spiritual longing becomes a byproduct of evolutionary pressure.

The concept of the image of God offers a different framework. It suggests that human intelligence is not accidental, but grounded in a deeper source. The ability to reason, create, and reflect on ourselves points beyond the biological mechanisms that support those abilities.

This does not deny the importance of the brain, which provides the structure through which thought occurs. But creativity, moral awareness, and self-reflection suggests that human intelligence may not be fully explained by structure alone.

If that statement is true, then the question of artificial intelligence becomes more complex. Creating systems that process information may not be the same as creating beings that possess creativity, moral awareness, spirituality, and self-reflection. These qualities define what it means to be human. They may also point toward the origin of intelligence itself.

**Deep Dive Box: Imago Dei**

The phrase Imago Dei comes from the Latin translation of Genesis 1:27, which describes humanity as created in the image of God. The concept has been interpreted in different ways, but most traditions agree on several core implications.

**First, human beings possess inherent dignity.** This dignity does not depend on intelligence, strength, or social status. It arises from being image-bearers. Every person shares this status.

**Second, humans are capable of reason and moral judgment.** These capacities are understood as reflections of divine attributes. The ability to recognize truth and distinguish right from wrong is central to this view of human nature.

**Third, humans are relational beings.** The biblical narrative presents humanity as created for relationship, both with God and with one another. This relational dimension reflects the idea that intelligence is not isolated. It seeks connection.

**Finally, the concept of *"Imago Dei"* suggests purpose.** Human life is not merely the result of physical processes. It is oriented toward understanding, growth, and relationship with its source.

These ideas shape how human intelligence is understood. Intelligence is not simply the ability to compute. It includes awareness, meaning, and the search for origin.

### The Bright Side: AI can extend human creativity without replacing the creator

A 2025 study[1] in Frontiers in Computer Science found that human-AI co-creative design can enhance creative performance, especially when the human remains the directing agent in the process. Used well, AI can function like an instrument that expands options, accelerates iteration, and helps people explore ideas they might not have reached as quickly on their own. Tools can support human imagination, but, the value still rests in the person using the system. The machine assists. The human imagines, judges, and chooses.

### Proceed with Caution: When a tool starts replacing reflection, something human is lost

Recent 2025 research[2] on generative AI and critical thinking found a pattern of concern: heavier dependence on AI was associated with lower critical thinking. A 2025 survey[3] of 319 knowledge workers found that higher confidence in generative AI was associated with less perceived enactment of critical thinking, while higher confidence in one's own ability was associated with more critical thinking. A system can help produce text, organize information, or suggest answers, but if people begin outsourcing judgment itself, they risk weakening part of what makes human intelligence distinct. Efficiency is not the same as wisdom. Fast output is not the same as moral or reflective thought.

## 1.7 THE OLDEST TEMPTATION

**In short:** The limits of being human have been tested since the beginning of time. The following narrations is based on events reported in the Genesis chapter of the Bible.

The garden was quiet. Light shone through the leaves, and the air carried no threat. Everything had been given with open hands. Only one boundary had been set, around a simple tree, and its un-assuming fruits.

The serpent slithered near the woman. He spoke as if he were helping her think more clearly. "I don't mean to pry," he whispered. "Is that really what God said?"

The simple question invited her to look at her gifts with suspicion instead of gratitude. She answered simply. "God said we are free. Only one tree is forbidden. We cannot eat its fruit, or we will die."

"Therefore, my dear, are you *really* free? Why such a limit? You will *not* surely die," the serpent pressed, with the calm confidence of someone offering a profitable secret. "If you take it, your eyes will open. He knows you will become like Him, knowing what He knows."

The serpent's words didn't change the fruit, but the story

around it. What once stood as a sign of trust between God and His creation, became in Eve's mind something withheld. A disturbing limit. The accuser had planted in her mind the thought that God was keeping greatness for Himself, that obedience was naïveté. That wisdom could be seized, instead of received.

That is how ruin entered our world. Not by force, but by suggestion, questioning the word of God. A suspicion against His goodness. A whisper that said the Creator was holding back, and that life could be increased or surpassed by taking His place. The story in Genesis talks about limits. First, a boundary is set. Then the accuser offers a suggestion designed to break the boundaries, but framed as a certainty: "You will be like God." The implication is that humanity is capable of becoming more than it currently is. That the limitation imposed by God is only temporary. The boundary is unnecessary. The path forward is self-directed transformation. The temptation is not disobedience but transcendence.

Today, this idea is becoming more familiar in all aspects of society. Human beings have always sought to move beyond their limits. We try to extend life, increase knowledge, and overcome weakness. We build tools to enhance strength, create medicine to resist disease, and develop systems to expand memory and communication. These efforts reflect the human drive to understand and improve the world.

The difficulty arises when *improvement shifts into transformation*. There is a difference between caring for what exists, and attempting to redesign it. **Stewardship seeks to preserve and cultivate. Transformation seeks to replace.** Sometimes, the line between the two can become difficult to see.

The language of modern technology often describes advances as upgrades, and human limitations as defects. Aging becomes a problem to eliminate. Memory becomes something to expand artificially. Intelligence becomes something that can be engineered.

In a purely technical sense, the assumption behind these ambitions is that humanity is an unfinished product. With the right tools, it can be improved. The goal is not simply to heal or support, but to surpass what currently exists. This reflects how humans understand themselves. On one hand, we recognize that we are finite, vulnerable, and dependent on the world around us. On the other hand, we seek to overcome those limits, and become self-sufficient and self-determining.

The narrative in Genesis captures this tension in symbolic form. The promise is not simply knowledge, but autonomy. To be "like God" is to define one's own nature, determine what is good according to personal algorithms, and move beyond restrictions that are felt as arbitrary.

Modern technology is often similarly motivated by a goal not only to understand the human mind, but to enhance it. Not only to treat disease, but to eliminate our biological vulnerability. Not only to build intelligent tools, but to create a kind of intelligence that surpasses human ability.

These goals are frequently framed as progress, and in many ways, they are. Technology has reduced suffering, extended life, and expanded knowledge. However, when humanity seeks to transcend its own nature, it implicitly assumes that human nature is something that must be improved, according to standards like efficiency, speed, or capacity. Intelligence becomes measurable in terms of processing power. Memory becomes measurable in storage. Decision-making becomes measurable in the optimization of results.

These metrics capture certain aspects of intelligence, but they do not capture what it means to be human. Creativity, moral awareness, spirituality, and self-reflection do not fit into technical categories. They resist simple measurement. They point toward a dimension of intelligence that is not defined by performance alone.

This is where the **distinction between pride and stewardship** becomes important. Stewardship recognizes limits and works within them. It seeks to understand, preserve, and responsibly cultivate what exists. Pride seeks to overcome limits entirely. It assumes that human design can be replaced by human invention. Stewardship asks how technology can serve humanity. Pride asks to redesign humanity through technology.

The temptation described in Genesis predates modern science, yet it speaks directly to this distinction. The promise of becoming "like God" is the promise of self-definition. Technology did not create this impulse. It only amplified it.

The desire to transcend limits is older than any machine. It appears in myth, philosophy, and religion long before the modern era. Stories across cultures describe attempts to gain immortality, perfect knowledge, or divine status. These narratives reflect a persistent human intuition that limitation is something to overcome. Modern technology provides new tools for pursuing that intuition. Artificial intelligence, genetic engineering, and neural interfaces extend the scope of what seems possible. The methods have changed, but the ambition remains the same.

This suggests that the central question is at the root of what it means to be human. What is humanity, and what does it mean to improve it? If human intelligence includes creativity, moral awareness, and the search for meaning, then improvement cannot be measured solely by increased capability.

The oldest temptation invites humanity to redefine itself. The challenge is to discern whether that redefinition reflects growth or departure. Technology can extend human ability. It can also reshape how humans understand themselves.

The distinction matters because it shapes the direction of innovation. If humans are merely biological machines, then redesign becomes logical, even mandatory. If humans are image-bearers, then the goal shifts drastically from surpassing humanity to understanding it more deeply.

The temptation to transcend limits remains powerful. It promises freedom, capability, and autonomy. Yet it also raises a question that continues to echo through the history of thought: what happens when humanity seeks to become more than human before it understands what being human means?

**Deep Dive Box: Cellular reprogramming and the anti-aging frontier**

In early 2026, the FDA cleared the first human trial of a cellular rejuvenation therapy[1] using partial epigenetic reprogramming. It was developed by Life Biosciences for age-related optic neuropathies. The immediate target is medical, but the broader significance is larger. Cellular reprogramming is part of a field driven by the hope that aging itself can be modified at the cellular level. The original purpose may be therapeutic, yet the cultural imagination around the field often reaches toward something bigger: not only healing decline, but rewriting the conditions of human limitation.

**The Bright Side: Soft auditory brainstem implants**[2] are designed for people with depressed hearing who are not eligible for cochlear implants. In macaque studies reported in 2025, the device produced high-resolution auditory percepts, and points toward better rehabilitation for profound hearing loss.

**Proceed with Caution: A beanie designed to turn thought into text**

On April 16, 2026 Wired[3] reported on Sabi, a startup developing a noninvasive brain-computer interface in the form of a beanie designed to decode internal speech into text. The device is aimed at convenience and speed, but it also represents a deeper shift. Thought is being treated as a stream that can be captured, digitized, and optimized for output. That makes it a vivid example of the modern urge to push past ordinary human limits through technology. The appeal is obvious, but so is the question it raises. When inner mental life becomes data, the issue is no longer only increasing human capability. It extends into privacy, agency, and what parts of being human should remain more than a technical surface to be mined.

# PART TWO

*The dream of becoming more than human*

Most people who eat a tomato in January think nothing of it. The person who owns the patent for that tomato, who decided what pests it would resist and what it would become, remains hidden. The soldier fitted with a neural implant makes the news for a day. The wheat engineered to survive a herbicide that kills everything around it gets a paragraph in an agriculture journal, or mentioned in passing by a food influencer. The pastor of the mega-congregation two zip codes from your church, the one who reads Genesis as a launching pad poetic tale rather than a boundary, skyrockets to fame and fortune and gets the book deals. These all belong to the same conversation. The dream of becoming more than human runs through all of them. It is older than patents, older than the word transhumanism. What changes across centuries is the sophistication of the tools. From the first quiet genetic interventions in the food supply, to the loud and confident transhumanist vision of an upgraded humanity moving past its biological ceiling.

This is a debate the church has not yet concluded: whether God intended human nature as a gift to be received, or a starting point to be exceeded. Serious theologians stand on both sides of that line. The upgrade dream is already here. The only question left is whether the church has anything to say about where it ends.

## 2.1 THE BIRTH OF THE UPGRADE DREAM

**In short:** transhumanism grew out of Enlightenment confidence in reason, evolutionary ideas about change, and technological optimism. Human nature itself came to be seen as something open to redesign. It is not only a technical movement but a philosophical one, because it depends on what we believe human beings are and whether we are meant to be upgraded or received as image-bearers of God. The following narration is loosely based on actual events documenting the use of brain to computer interfaces.

Elena stood across the street from the Ascend Institute and watched her own reflection in the glass panes. The building rose in glistening rectangles above the city. At the entrance, a vertical ad played a slow sequence of faces. A child reading. A surgeon in an operating room. An old man smiling at a granddaughter. She had seen this before. The advertisement oozed the smooth confidence that their product had the same finality as electricity. Like antibiotics. Like computers.

Elena flicked the edge of the visitor badge clipped to her coat, and crossed the street. Inside, the lobby smelled faintly of citrus and cheap floor polish. A receptionist in a gray suit greeted her by name before she could say anything.

"Hello, Dr. Anesh. Adrian is expecting you," the woman said.

Elena almost smiled. She was not a doctor. She had corrected people before, back when titles still seemed worth defending. Historian, she would say. Research fellow. Writer. Observer. The elevator opened onto the twelfth floor.

More glass walls framed a long corridor lined with holographic diagrams of the brain. Neural networks lit up in blue and gold, brightening and dimming in careful sequences. Under each image ran a caption in the language she had begun to hate: Memory optimization. Emotional regulation. Cognitive acceleration. Performance expansion.

She followed the corridor into a conference room where Adrian Vale stood alone, hands folded behind his back, looking over the city.

He turned as she entered."Elena, so nice to see you. Thank you for coming."

He looked young. No white hair. No grandfatherly gravity. He looked like a man who had learned to make poise as part of his image, but it didn't really fit him.

"You said I could speak to the first subjects alone," Elena said. She wasn't about to waste her time with the CEO's pep talk.

"And you will." His smile was carefully elegant. "We have nothing to hide."

That was the phrase people used when they did, and expected nobody to find out.

He gestured, and the wall behind him shifted to a rendering of the implant. A thin crescent, no larger than a fingernail. The rotating, expanded image made it look no more ominous than a child's toy.

"We are well past theory now," he said. "The device integrates neural networks across sensory association regions."

"And do what? From what I understand, it changes the personality of the recipient," Elena said.

"It does not replace the person, just removes bottlenecks. Human potential has always been boxed in by biology. Slow recall. Emotional distortion. Cognitive fatigue. Nexis reduces those limits," he chuckled. "Well, more like it blows them out of the water!"

"Reduces them into what?"Elena asked, looking at the image.

He studied her for a moment, as if deciding whether her skepticism deserved patience."Into freedom," he said.

She opened her notebook, though she was not planning to write. There it was. Not medicine. Not treatment. Not assistance. This guy was selling freedom. But freedom from what? "You're not treating disease," she said. "You're not curing anything."

"Not in the traditional sense," replied Vale.

"Is there any other? Being human is not a defect."

Vale laughed softly, without malice."That's the historian in you," he replied. "We have always used tools to overcome limitation. Glasses correct vision. Computers extend the ability to calculate, and memory. Nexis is not any different. It is just the next sentence in human history."

"Tools used to stay outside the body," she replied.

"That distinction was artificial. Pacemakers are inside the body. What about artificial joints? Cochlear implants?" He stepped closer to the display, and touched the glowing image with one finger. "Your resistance to progress is emotional, not rational."

Elena did not answer. She had heard versions of this same pitch for months. The old confidence and enthusiasm of the Enlightenment, now dressed in new fabric.

He offered her a schedule and a list of names, with the polished assurance of a man who believed the future had already consented to his plan.

The first subject was a teacher named Miriam Chen. She sat in a quiet interview room with a glass of water and perfect posture. Nothing about her looked altered. Mid-forties, black sweater, no visible scars.

"How do you feel?" Elena asked, unsure as to how to start. How does one interview a computer?

Miriam considered the question carefully for a few moments. "Clear," she said. "That's the first word. Clear in a way I didn't realize was possible."

She described reading two books at once and retaining them completely. She described listening to a conversation in another room while grading papers and drafting a lecture in her head. She described *all* her memories existing not as fragments now, but as indexed continuity. Available at will.

Elena leaned forward. "That sounds wonderful. But what about emotions?"

Miriam smiled as though she had expected that one. "Less burdened. Lighter."

"Can you explain that?" Elena asked.

"I still *feel,* just like before," Miriam said. "I just don't get overwhelmed. When my son called last week and told me he was moving to Berlin, I felt sadness. I also felt pride, concern, love. I had practical questions. All of it in order. Nothing flooded my mind, or trapped me in confusion."

"And that felt better?" Elena pressed.

"Yes. I felt in control."

The second subject was a composer named Daniel Ortiz.

He was enthusiastic in the way of new converts.

"I hear structures now," he said. "Not just melodies. Full architectures. I can hold an entire symphony in my mind. Revise it, test alternative developments. Before I ever touch a keyboard."He laughed and shook his head. "It used to take me weeks to find a transition. Now I can hear twelve versions and discard eleven in an hour."

"That sounds useful," Elena said.

"It's extraordinary."

He spoke for twenty minutes about scale, harmonic compression, rhythmic recursion. Elena let him go on. Then she asked the question that mattered.

"Has it changed what music feels like?"

He paused. The room went quiet enough for her to hear the vent moving air above them.

"I don't know," he said at last, looking down at his hands. "Music used to surprise me," he continued. "Sometimes a melody would appear in my mind. Like I had found it, not made it. That doesn't happen now. My work is better. Much better. But it feels..." He searched for the word. "It feels intentional."

By the fourth interview, Elena had begun to notice the pattern. The subjects were sharper, steadier, more controlled. They were also harder to read. Not colder exactly. More sealed. As if some of the irregularities of ordinary life had been smoothed away.

One man remembered every conversation with his dead wife, down to pauses and inflection. Yet when Elena asked which memory he returned to most often, he answered like an archivist selecting files.

"A morning in Lisbon," he said. "Café on Rua Augusta. High emotional density. Strong visual."

She waited for more. He nodded once, thinking that perhaps he had answered.

When Elena went home that night, she did not turn on the lights.

Her apartment overlooked the river. She sat at the kitchen table with her coat still on, and opened an old storage box she had not touched in years. Inside was the usual treasure trove of memories. Notebooks from graduate school, photographs, folded copies of articles she had once thought important enough to mark by hand.

At the bottom lay a page from her father's handwriting. He had copied a passage from Francis Bacon for her when she began her first fellowship, some solemn encouragement about knowledge and humility. She remembered the day because they had argued afterward in the campus parking lot.

He had died six months later, before either of them could apologize. Elena sat with the page in her hands. She could not remember every word he had said that afternoon. She could not reconstruct the exact angle of the light, or the order of his gestures. What she remembered was the pain of having been loved by someone stubborn enough to ignore her feelings. She thought of the man in the interview room, calling Lisbon *high emotional density*.

On the third day, Vale invited her into the procedure wing.

Through a glass partition, she saw a row of private recovery rooms. In one of them, a girl no older than twenty sat upright in bed, touching the bandage behind her ear while a nurse spoke to her.

"We're still selecting carefully," Vale explained. "High-functioning candidates with stable psychological profiles.

People who understand the opportunity."

"Opportunity for what?"

He looked at her, and for the first time his practiced gentleness dropped away.

"For the end of helplessness," he replied arrogantly.

Elena turned from the glass."Not all limitations are bad."

"I guess," he said. "Some of them may be glorious in poetry. I find them intolerable in practice."

He led her into his office. Books lined one wall, though she suspected they had been chosen for texture as much as content. Darwin. Descartes. Kurzweil. Haraway. Augustine, placed a little too visibly.

Vale poured water into two glasses but did not offer one.

"You think I despise the human person," he said.

"I think you believe the human person is a draft."

"And that's insulting to you?"

"I think it is dangerous."

He nodded, almost pleased.

"Let me say it plainly then," he began. "Humanity has always been transitional. Fire, agriculture, writing all changed us. Industry changed us. Computers changed us. The internet changed us. AI changed us. Every age redraws the shape of the human. We're the species that escapes itself."

He leaned forward."Why would evolution stop with us?" he continued. "Why would intelligence refuse the means to improve itself? Why would a rational creature worship its own limitations?"

There was no arrogance in his voice. That was what troubled her. He sounded like a man reciting weather.

"Because some limits teach us what kind of creature we are," Elena replied.

"That is a beautiful sentence," he said. "It's also what every age says, just before it moved pass them."

When she left the institute, she walked without direction for nearly an hour, and stopped at a small public square near the river. The afternoon had blossomed into one of those clean city days that made every surface look new. A young mother sat on a bench while a little boy in a red coat tried to jump from one paving stone to another without touching the seams. He failed repeatedly. Each time, the same burst of laughter, pure, and useless, and complete. Not efficient or optimized. Not converging toward measurable improvement. Just delight.

The mother clapped for him as though he had crossed an ocean.

Elena stood very still.

The city around her was full of things built to defeat weakness. Bridges, hospitals, data centers, prosthetics, levees, code. She was not naïve. She did not despise progress or medicine. She didn't romanticize suffering. Yet all at once it seemed to her that the question was not whether people would gain power. Of course they would. The question was who would decide which gains were worthy.

In her left coat pocket lay the invitation Vale had slipped to her before she left: *Eligibility consultation scheduled.*

A courtesy, he had called it. No pressure. An observer should understand the thing from the inside.

She took the card out and looked at it. White stock. Silver lettering. A future presented as administrative routine.

Across the square, the boy had given up on the game and was now crouched beside a pigeon, speaking to it with solemn intensity. The bird tilted its head as if considering the terms.

Elena thought of all the words she had heard inside Ascend. Upgrade. Enhancement. Performance. Acceleration.

None of them were false. That was the difficulty.

She folded the card once, then again.

The temptation of her age, she realized, was refinement. It was the promise that the deepest questions of identity could be answered by increasing our capacity like one expanded a computer's memory. It was the old voice from Scripture rephrased in modern grammar, whispering that to move beyond what was given was to become fully free.

She stood there a moment longer, watching the child, the pigeon, the mother, the late sun over the river. She tore the card in half, and dropped it into a trash bin beside the bench.

This fictional story illustrates the gradual emergence of the idea that humanity might become more than human, through shifts in how people understood reason, progress, and technology. Over time, these ideas formed a framework in which human nature itself could be seen as something open to redesign.

One of the earliest influences came from **Enlightenment Rationalism.** During the historical period of the Enlightenment, thinkers emphasized the power of human reason to understand and shape the world. Nature was no longer seen only as something to observe. It became something to analyze, measure, and ultimately control. Scientific discovery accelerated, and confidence in human intellect grew. The universe appeared increasingly ordered and comprehensible, governed by laws that could be discovered and applied.

This shift changed how humanity viewed itself. If reason could uncover the principles governing nature, then those principles might also be used to improve human life. Progress changed from a hope to an expectation. Knowledge was expected to accumulate, each generation inheriting a more advanced understanding than the one before. This confidence in reason created a narrative in which humanity was not fixed but constantly moving forward of its own means.

The idea gained further momentum with the rise of **Evolutionary Theory**. The Darwinian account of natural selection described life as a process unfolding over time. Species change. Traits emerge. Adaptation shapes survival. Humanity itself became part of this long developmental story.

Once human beings were placed within an evolutionary framework, a new possibility appeared. If evolution had shaped humanity, then humanity might continue to evolve. The process did not have to stop. In fact, it could be accelerated. It should be accelerated.

Evolution was no longer only something that happened to humanity. It became something humanity might be able to direct. ***Natural selection could be supplemented by intentional design.*** The future of the species could be shaped deliberately rather than passively. This idea merged naturally with technological optimism. The industrial age demonstrated that machines could extend human strength. Electricity extended human reach. Computers extended memory and calculation. Each innovation seemed to confirm the same pattern. Technology allowed humanity to overcome limitations that once appeared permanent.

Over time, these strands began to converge. Reason suggested that nature could be understood. Evolution suggested that humanity was not fixed. Technology suggested that limits could be overcome. Together, they formed a powerful narrative: humanity was a transitional stage, not a final form.

From this narrative emerged an explicit ideology that came to be known as transhumanism.

***Transhumanism is the belief that human beings can, and should, move beyond their biological limitations through technology.*** The goal is not simply to treat disease or extend comfort. It is to enhance intelligence, extend lifespan, and reshape human capabilities. In some versions, the ambition goes further. **Humanity itself becomes a starting point for something new.**

Within this framework, the human body is seen as modifiable. The brain becomes a system that might be augmented. Memory might be expanded, perception enhanced, and cognition accelerated. Artificial intelligence could merge with biological intelligence. Genetic engineering could alter traits before birth. Neural interfaces could connect minds to machines. These possibilities are often accepted as logical extensions of existing trends. Medicine already replaces failing organs. Computers already assist decision-making. Algorithms already outperform humans in certain tasks. The next step appears to be integration. Human intelligence and technological capability become increasingly intertwined. The language used to describe this future reflects common terminology used in engineering. Humans are discussed in terms of upgrades, optimization, and performance. Limitations become inefficiencies. Aging becomes a defect. Memory becomes storage. Intelligence becomes processing power.

This perspective shifts the meaning of progress. Improvement is no longer limited to tools that serve humanity. Instead, humanity becomes the object of improvement. The distinction between tool and user begins to blur. At this point, the dream of becoming more than human takes shape. It is not framed as rebellion, but presented as advancement. The goal is to overcome weakness, extend knowledge, and increase capability. Who could be against any of these? In this sense, transhumanism appears as a continuation of scientific progress.

Yet the idea also raises questions about identity. If intelligence is enhanced beyond natural limits, is the result still human? And, why should we attempt to expand intelligence, when we have such a hard time defining it? If memory is expanded artificially, does personal identity change? If cognition is merged with machines, where does the person end and the technology begin?

These questions reveal that transhumanism is not only technological. It is philosophical. It depends on assumptions about what humans are and what they should become. If humans are biological machines, then redesign appears natural.

The dream of becoming more than human rests on a deeper claim. It assumes that human nature is incomplete. Given enough knowledge and engineering, it can be improved. This assumption reflects the convergence of

Enlightenment Rationalism, evolutionary thinking, and technological optimism. Reason promises understanding. Evolution suggests change. Technology offers the means.

**Deep Dive Box: Key Transhumanist Thinkers**

The biologist Julian Huxley[1] introduced the term "transhumanism" in the mid-twentieth century. He described humanity as capable of transcending itself through science and technology. In his view, evolution had produced human intelligence, and that intelligence could now guide further development.

Later writers expanded the concept. The philosopher Nick Bostrom explored the ethical and technological implications of human enhancement. He argued that advances in biotechnology, AI, and neuroscience could fundamentally transform human capabilities. His work framed transhumanism as a serious philosophical and scientific discussion rather than speculative fiction.

The inventor and futurist Ray Kurzweil popularized the idea that technological growth follows an accelerating curve. He proposed that AI and human cognition might eventually merge, leading to a new stage of intelligence. This moment, often called the *singularity,* represents the point at which technological progress transforms human life in unpredictable ways.

These thinkers differ in emphasis, but they share a common assumption. Human nature is not fixed. It can be reshaped through knowledge and engineering. The future of intelligence may extend beyond biology.

This vision captures the essence of the upgrade dream.

**The Bright Side: Technology has genuinely extended human ability**

Machines have extended strength, electricity has extended reach, and computers have expanded memory and calculation. Medicine replaces failing organs. Algorithms assist decisions. These developments have relieved suffering and widened what human beings can do. Many technologies do serve human life in meaningful ways, and that service should be recognized clearly.

**Proceed with Caution: When progress turns the person into a project**

The danger begins when humanity itself becomes the object of optimization. Once the body is treated as modifiable hardware, and the mind as a system to upgrade, the distinction between tool and user begins to fade. Memory becomes storage. Intelligence becomes processing power. Aging becomes a defect to be eliminated. At that point, the question is no longer how technology can help human beings live well. The question becomes what kind of being will remain after human identity is recreated through engineering.

## 2.2 CROPS TO GENES

**In short:** From breeding crops to editing genes: what is the Christian responsibility in being good stewards of creation? The following narration is based on actual events. Names and brands have been altered for privacy reasons.

The first letter came in July. Folded sharply. Machine-made. Eli Turner stood at the kitchen counter, the envelope open in both hands. The coffee beside him went cold. The return address carried the name of one of those Big Ag corporations. Everyone in the county knew it. Their seed signs stood along the highways. Their field reps wore pressed shirts and drove trucks washed twice a week. His wife, Nora, looked up from the sink.

"What is it?"she asked.

Eli read to the bottom, then started again, slower this time, as if a second pass might change the words.

"They say I planted their soy," he replied, incredulously.

Nora dried her hands and came to stand beside him. He showed her the page. The language was calm. Their inspectors had identified proprietary genetic traits in crop samples taken from acreage under his management. Unauthorized possession of blah blah blah. Patent infringement. Immediate compliance required. Records requested. Settlement options available.

"You never bought from them?" she asked.

"No."

"Not ever?"

"You kidding? Not once."

The old clock above the pantry clicked into the silence. Through the window, the west field lay flat beneath the morning light, green and orderly rows running toward the tree line. It looked like the same field Eli had worked for thirty years. Soybeans. Dirt. Heat rising by noon. Work that started before dawn, and ended when there was no light left. He had saved his own seed and planted it that spring, as he had done for decades. He knew what was in his ground, measured it in cash, diesel, and sleep.

He called the number at the bottom of the letter.

A woman with a bright, trained voice kept him on hold for ten minutes, transferred him twice. Then a man answered. He introduced himself, last name and title. He said he was from the compliance division.

"There must be some mistake," Eli said. "I didn't plant your seed."

"Our test results indicate the presence YMax 4 in your field," the man replied.

"I'm telling you I didn't buy your product," Eli exclaimed.

"That is one possibility we are prepared to review during the documentation phase," the voice informed.

Eli gripped the phone tighter."One possibility? What's the other one?"

"Unauthorized use, obviously," the man replied.

Eli looked out the window at the field again. Somewhere beyond the cottonwoods a tractor was already moving, its engine low and steady across the morning. He said, slower now, "My place borders three farms that use your seed."

"We're aware of neighboring operations."

"Then seed could have drifted," Eli explained.

"Our patents protect our interests, they do not excuse infringement."

Eli didn't reply.

The man continued, still courteous in tone. "If you cooperate, these matters can often be resolved efficiently."

Efficiently. The word landed with the dull weight of money leaving a bank account.

By noon the news had reached the grain elevator, and by evening three neighbors had called. One spoke in sympathy. Another warned. One advised him to settle quickly. Don't mess with them. Don't stick to principle against a legal department.

"They do this hard," the neighbor said. "They don't have to prove what you think they have to prove. They just up spend you. Make it cost more to fight than to kneel."

Eli thanked him and hung up.

That night he walked the perimeter of the north field with a flashlight. Nothing to see by then, except fence posts, ditches, insects in the weeds. He knew where the neighbor's farm began, where the drainage cut across the back corner, where the birds settled in late afternoon, and lifted in a rush when the combine came through. The wind crossed property lines without asking, and so did pollen. So did seeds in the tread of a truck tire or the belly of a blackbird. Nature had never respected the legal imagination of corporations. It scattered where it would. Yet, now a company was speaking as if they owned the wind.

A week later two inspectors arrived.

They wore clean boots, and carried tablets. One was young enough to look embarrassed. The other had the dry, neutral face of a man who had decided the outcome of that visit long ago.

Eli met them by the equipment shed. "You got a warrant?"

"We're here to verify inventory, and discuss voluntary sampling," they answered.

"You already sampled."

"We require additional confirmation," one of the men added.

"No."

The older man nodded as though refusal fit neatly into a chart somewhere."I should let you know," he said, "that noncooperation may be considered in subsequent proceedings."

"Subsequent proceedings?" Eli repeated. "You mean, when you sue me? For what blew into my dirt?"

The younger one looked down at his tablet.

The older man said, "The patented traits were found in your crop."

"I didn't ask for them," Eli responded.

"That does not change the presence of the technology in your property."

Eli laughed once, without humor. "Technology! You mean beans?"

The man did not answer. They left after ten minutes. The dust from their SUV drifted over the lane long after the vehicle was gone.

In September the suit came.

The papers named Eli in a county case that felt too large for a county courthouse. Patent infringement. Unauthorized cultivation. Rain had come late that year. Fuel prices had climbed. One gearbox on the combine went out in August. He hadn't grown much but debt. Now a company with annual revenue larger than the economy of a whole country was asking the law to treat him as a thief.

At church, people lowered their voices when he passed.

At the diner, men stopped talking when he entered, then started again too loudly. Everyone knew the story even if they did not know the facts. Some assumed he had done it, and got caught. Others assumed he had been marked because he was too stubborn to settle quietly.

The hearing was set for early November.

The courthouse smelled of old paper and wood paneling. Eli sat beside his lawyer, a local man who had spent most of his career on land disputes, probate fights, and one ugly case involving a grain contract gone sideways. He had warned Eli three times already that fairness wasn't the issue here.

Across the aisle sat the company team. Three lawyers in navy suits, two boxes of files, one expert witness. They looked rested and prepared. Professionally grave, not angry. Anger belonged to people with something to lose.

When the arguments began, Eli tried to follow the legal language. The company's counsel spoke of rights, trait prevalence, market protections. He spoke as if seeds were no longer seeds but a licensed delivery system. They assumed the old habits of farming had simply failed to keep pace with progress.

Eli stared at his own hands.

Those hands had lifted rocks from fields, repaired broken couplings, buried a dog behind the barn, caught his daughter when she jumped from the hayloft at nine. They had planted what he could afford, and harvested what the weather spared. They had never signed a contract with the wind.

When his lawyer stood, the room lost some of its polish. He spoke plainly. Cross-pollination happens. Seed drifts. Birds carry. Trucks spill. Ditches flood. No evidence showed intentional planting. There were no purchase records, licensing agreement, or rational benefit. Only the presence of some of their seed on Eli's land.

Presence, Eli thought. That was all. Not intent. Just presence.

After lunch the company called its expert, who explained trait detection with the confidence of a man whose work entered courtrooms before. The patented gene sequence was present in Eli's crops. The markers were unambiguous. The field contained protected material. Nobody could dispute the science. That was the worst part.

Outside, while the court recessed, Eli stood on the courthouse steps beneath a white sky. Cars moved through town as they always had. A teenager in a red apron carried out a bucket of mop water from the café, and dumped it into the gutter. Life continued with the rude indifference of a Tuesday.

Nora stood beside him."You all right?"she asked.

He took longer to answer than he meant to. "They own the seed," he said. "I know that's how the law sees it. But they're talking like they own the ground too."

Nora said nothing.

He looked out at Main Street, then beyond it in his mind to the fields, to the stubble and drainage and the places where property lines vanished under weather.

"It's not even just the ground," he said. "They want water to stay put for them? Birds don't ask permission."

She touched his arm.

Back inside, the settlement offer arrived before the ruling. Reduced damages if he agreed to future compliance, licensing review, and inspection access. No admission of wrongdoing required, which was the company's version of mercy. He read the paper once, then set it down. His lawyer leaned back in the chair and rubbed his forehead. "If you fight and lose, it gets worse."

"If I sign, what does that make it?"

The lawyer did not answer.

That night Eli walked to the edge of the south field, and stood in the dark. No flashlight this time. The air carried the smell of turned soil, and distant rain. He thought of his father, dead eight years now. He used to cup seed in one palm, and speak of it like a promise. You put life in the ground, he would say. You work. You wait. You pray. That had been farming as Eli understood it. No season guaranteed. Now men in suits wanted to tell him that life could be patented. He bent down, scooped a handful of dirt, and let it fall back through his fingers.

The next morning he told his lawyer not to settle.

It was not courage, exactly. What he felt was more stubborn, and more tired than that. He refused to speak a lie because it was well packaged. He wasn't going to call drifted seed an act of theft. He wouldn't let paperwork rename what wind had done.

The case could cost him the farm. He knew that now. But as the sun rose over the field, he understood something with painful clarity. The fight was no longer only about beans, or patents, or one bad season. It was about ownership, and he wasn't for sale.

Behind him, the house door opened. Nora stepped onto the porch with two mugs in her hands.

Eli turned and took one, leaving the field at his back, the papers waiting on the table, and the wind blowing where no fence could hold it.

For most of human history, crop improvement meant selection of better traits.

Farmers chose the strongest plants, and saved their seeds. Shepherds bred animals that were healthier, faster, or more productive. Over generations, these choices shaped crops and livestock into more suitable investments. Wheat became more resilient. Dogs developed specialized traits. Cattle produced more milk. The process was slow, but simple in principle. Humans worked with variation that already existed, and guided the outcome through selection. Nothing fundamental about the organism was rewritten. Traits were combined, emphasized, or reduced. The underlying genetic code remained untouched.

Today, that boundary has been moved.

Modern biotechnology allows scientists to edit DNA directly. Instead of selecting traits across generations, researchers can alter the genetic sequence itself. The change does not require breeding. It can occur within a single organism, sometimes within a single cell. This transition marks a significant change in how humanity interacts with life. Selection works with what is available. Editing introduces new variation intentionally. The difference is subtle in language but substantial in practice. Humanity has moved from guiding evolution to intervening in the code that underlies it.

One of the most discussed tools in this shift is CRISPR

(which stands for Clustered Regularly Interspaced Short Palindromic Repeats). The name refers to a technique derived from a natural defense mechanism found in bacteria. Scientists discovered that certain microbes use CRISPR systems to recognize and cut foreign genetic material. Researchers adapted this mechanism into a precise method for editing DNA.

With CRISPR, a targeted sequence in the genome can be located and modified. A gene can be removed, replaced, or altered. The process is still developing, but it has already been used extensively in research and medicine since 2012. Experiments have focused on correcting genetic disorders, modifying immune responses, and studying the function of specific genes.

The technology is powerful because it operates at the level of information. DNA functions as a biological instruction set. Editing that instruction set changes how cells develop and function. In principle, this allows direct control over traits that once required generations of selection.

Genetic engineering extends this idea further. Scientists can introduce genes from one organism into another, or even produce synthetic versions of desired traits. Crops have been engineered to resist pests or tolerate harsh conditions. Bacteria have been modified to produce insulin. These applications often focus on solving practical problems, but they also demonstrate the broader possibility of constructing organisms for commercial purposes.

As these capabilities expand, the idea of "designer organisms" becomes more acceptable. The term refers to organisms whose traits are selected and engineered before development. In agriculture, this may involve improving yield or resilience. In medicine, it may involve correcting inherited conditions. In more speculative discussions, the concept includes expanding human traits such as intelligence, physical ability, or resistance to disease.

These developments continue the pattern introduced by selective breeding, but they operate at a different level. Selection works through variation over time. Editing works directly on the instructions themselves. The timeline shortens and the apparent degree of control increases.

This shift contributes to the broader idea that biology is programmable. If traits can be edited, then improvement appears to be an engineering problem to be solved. The same language used for software begins to appear in discussions of life. Genes are described as code. Mutations are described as software bugs. Editing becomes debugging.

The analogy is appealing, but it raises questions about complexity. Biological systems are interconnected. A single gene often influences multiple traits. Changing one part of the system may, and often does, produce effects elsewhere. The genome functions less like a simple program and more like a dynamic network. It is currently impossible to predict the consequences of a single genetic modification on the full spectrum of traits.

These complexities shape the debate around long-term genetic change. One perspective emphasizes that mutations and modifications accumulate over time, allowing evolution to produce increasing complexity. Another perspective raises concerns about degradation. Some researchers argue that most mutations are neutral or harmful, rather than beneficial. Over time, the accumulation of small negative changes may reduce overall genetic integrity.

This view is sometimes described as genetic entropy. The idea draws from information theory, suggesting that biological systems tend toward loss of functional information, not toward increase. Mutations introduce changes, but beneficial changes are rare. Harmful changes accumulate slowly, often remaining hidden for generations.

The argument does not deny that variation occurs. It suggests that even just maintaining complex systems requires more than random change, let alone increasing complexity. If true, this would debunk the assumption that genetic modification always leads to improvement. Editing DNA may correct specific problems, but long-term outcomes depend on interactions between genes that are not fully understood.

These debates highlight a broader issue. The ability to edit genes does not automatically provide a complete understanding of what genes do. The genome contains layers of regulation, interaction, and expression that remain under study. Changing one element may affect many others in subtle ways.

The move from breeding to editing reflects a deeper change in how humanity approaches biology. Selection assumes that life develops through gradual variation. Editing assumes that life can be engineered directly. The first approach works with complexity. The second attempts to redesign it.

This shift aligns with the broader theme of enhancement. If intelligence, health, and physical ability have genetic components, then altering those components appears to offer a path to improvement. The possibility contributes to the idea that humanity can move beyond its current limits. The transition from breeding crops to editing genes therefore represents more than a technological advance. It reflects a change in mindset. Life is no longer only cultivated. It is increasingly treated as modifiable at the level of its instructions.

When humanity gains the ability to rewrite the code of life, what does it mean to *improve* it?

**Deep Dive Box: He Jiankui and the line between treatment and redesign**

In 2018, Chinese scientist He Jiankui [1]announced that he had used CRISPR-Cas9 to edit human embryos[2], resulting in the birth of the first reported genome-edited babies. He had targeted a specific gene, CCR5, in an attempt to make the resulting children more resistant to HIV. The case was widely condemned, not only because of safety concerns, but because it crossed into heritable human germline editing without broad scientific support or ethical oversight. The children were not being treated for an existing illness. Their genomes were being altered at the embryonic stage in a way that could be passed to future generations. That distinction matters. This was not simply medicine correcting disease in a patient. It was a step toward redesigning human inheritance itself. The case exposed how quickly the language of prevention and improvement can move past therapy and into the engineering of future human traits.

**The Bright Side: Insulin-producing bacteria show the bright side of genetic engineering**

In the late 1970s, scientists inserted the human insulin gene into E. coli, allowing the bacteria to make insulin protein. This led to Humulin, developed by Genentech and Eli Lilly, which became the first FDA-approved recombinant DNA drug in 1982. One of the clearest positive examples in this field is the use of genetically modified bacteria to produce human insulin. Instead of relying on animal-derived sources, scientists can engineer microbes to manufacture insulin at scale for people with diabetes. The goal is not to redesign the human person or create a superior kind of life, but to meet a real medical need with greater precision and reliability.

**Proceed with Caution: Editing the code is not the same as understanding the whole system**

The complexity of the genome suggests caution. The same system that allows for change also resists simple redesign. Traits emerge from interactions across many genes and environmental influences. Editing one element may not produce predictable results.The chapter's main warning is that control at one level can create confidence that exceeds understanding. Genes do not operate in isolation. A single change may affect several traits, and the full range of consequences may not be visible at first. That makes the software analogy misleading if it is taken too far. Life is not a simple program with clean lines of code. It is a deeply interconnected system. Once biology is treated mainly as something to optimize, the risk grows that intervention will move faster than we can truly predict or control.

## 2.3 THE TRANSHUMANIST VISION

**In short:** The Upgrade Path: Digital consciousness, neural implants, life extension, and human-machine hybrids. The following is a fictional narration based on the current transhumanist trend.

Jack signed the waiver twice. Once on the tablet at the front desk, and again in a smaller consultation room, where the provider slid a second form across the table and tapped the signature line with one polished fingernail.

"The first covers standard consumer use," she said. "The second covers guided induction."

Jack looked down at the page. The language had that careful, modern tone that made everything sound both safe and inevitable. Sleep modulation. Neuroacoustic assistance. Noninvasive stimulation. Enhanced dream awareness. The product name sat at the top in thin silver letters: RING

He signed. The provider smiled professionally. "First session?"

"Yes."

"Any history of parasomnia, seizures, or panic disorder?"

"No."

She entered something on her screen. Jack watched the reflection of her hands in the glass wall behind her. Outside, the clinic looked more like a boutique tech showroom than a medical office. Matte white surfaces, indirect lighting, and a curved display showing animated brainwave patterns in pale blue. Three other clients sat in low chairs with the same strained composure people wore before cosmetic procedures or expensive dental work.

The provider closed the tablet."My name is Dr. Voss. I'll fit the device and monitor the first cycle."

"You're a doctor?"

"Neuroscience, not medicine."

He nodded as if that clarified something.

She stood and led him into the sleep room.

It was darker there, though not completely dark. A narrow bed stood against one wall. There was a chair, a shelf, and a wall monitor dimmed almost to black. On a tray sat the headband itself. RING looked lighter than he expected. A pale flexible band carried a thin crown of sensors and two recessed modules near the temples.

"Our device uses EEG and near-infrared sensing to detect sleep-stage transitions," Dr. Voss said as she lifted it carefully. "The guided mode adds timed stimulation when patterns consistent with REM onset appear."

Jack lay back while she adjusted the band around his head."And this helps me realize I'm dreaming?" he asked.

"That is the claim," she added.

"The claim?"

"We're careful with language here."

He almost laughed."That doesn't sound like the ad," he said.

She tightened one side of the band and stepped back. The monitor beside the bed flickered alive with soft waveforms. His name, heart rate, signal quality, and pre-sleep baseline appeared in cool light."Just rest," she said. "If you wake during the session, don't sit up too quickly.

Speak normally. The microphone is live." She moved toward the door, then paused. "One more thing. Lucidity can feel exhilarating the first time. It can also feel destabilizing. Try not to force control too early."

"Control of what?"

"The dream."

After she left, Jack stared at the ceiling for a long time.

He had not come for novelty alone, though novelty had played its part. The company's language about exploration and enhanced self-awareness had drawn him in the way these things always did, by mixing promise with vagueness. Beneath that was a duller truth. He had been dreaming about his brother for months. Not every night, but often enough to leave a residue in the day.

In those dreams, he saw Sam at seventeen, leaning against the fence behind their father's house, talking about motorcycles he could not afford. Sam at twenty-three, laughing too hard at some joke already forgotten. Sam in the hospital, thinner than Jack remembered, saying he was fine when nobody in the room believed him. The dreams never stayed long enough. They broke apart the moment Jack noticed them. Maybe RING would hold the moment in place.

The lights dimmed further.

He listened to the low hum of conditioned air, and tried not to think about being watched while he slept. At first there was only the slow disconnection of ordinary fatigue. His thoughts loosened. The edges of the room thinned. He drifted in fragments, half aware of his own breathing and half somewhere else. Then a faint pulsing sensation touched his temples, delicate enough to doubt. It was not pain. It felt more like someone tapping a glass far away.

The darkness changed.

Now, he was standing in the old field behind his childhood home. The grass moved in long waves with the late afternoon wind. The sky had that impossible color dreams borrowed from memory and improved. Beyond the field stood the rusted gate their father never fixed, and beyond that ran the road toward town.

Jack looked down at his hands.

For a second he forgot everything else and only stared. Skin. Knuckles. Dirt beneath a thumbnail. The details were too clear. Then the thought arrived with enough force to make the air feel sharper around him.

This is a dream.

The field brightened all at once, as though the world had turned toward him. His chest tightened. He remembered the provider's warning. Do not force control too early.

The wind passed through the grass again. Far off, a bird rose from the fence line and cut across the sky. Then he saw Sam. Seventeen, maybe twenty. One hand hooked into the back pocket of his jeans. He stood by the gate and looked toward Jack as if he had been waiting there for some time.

Jack did not move. Dream or not, the sight of Sam struck with blunt force. This was not symbolism or memory alone. It was the shock of seeing a dead brother look alive. Sam smiled first."You took your time," he said.

Jack walked toward him, then stopped halfway, afraid the dream would collapse if he crossed too quickly. "I know this isn't real," he said.

Sam shrugged. "That never stopped you before."

It was exactly the kind of answer his brother would have given. For a second, Jack hated the device for it. Neural induction, pattern prediction, and machine-guided memory had wrapped his grief in dialogue and fed it back as presence.

"You're not him," Jack said.

"Probably not," he replied.

Jack looked around. The gate. The road. The low hum of insects in the grass. Everything seemed held together by some intelligence that was not his, and yet came from nowhere except his own sleeping mind.

"What happens if I try to change this?" he asked.

Sam tilted his head. "Try it."

Jack turned and looked toward the house. It stood in the distance, white paint grayed by years, the porch sagging slightly as it always had. He fixed his attention on the porch light and tried to make it turn on. Nothing happened.

He tried again, harder this time, with the odd inward effort dreams seemed to require.

The porch light flared gold.

A strange joy moved through him. It was not because the light mattered. It was because his will had touched the scene, and the scene had answered.

When he looked back, Sam was grinning. "There you go."

Jack laughed then, short and disbelieving. The field wavered.

Somewhere beyond it, faint and thin, another sensation intruded. A voice. His name, distant as if underwater.

"Jack."

Dr. Voss.

The field darkened at the edges.

"Not yet," Jack said, though he did not know whether he was speaking to the dream or to the provider.

"Jack, you are entering a lighter stage. I need you to respond."

Now the words came through the dream itself, blended into the air. Sam glanced toward the sound as if he could hear it too. Jack felt anger rise, sudden and childish. He had paid for this. He had paid to step past the usual boundary where the mind lost hold and morning erased everything.

"Can I come back here?" he asked his brother. Sam looked at him for a long moment.

"You can come back to something," he said.

The answer annoyed him because it was the kind of answer people gave in dreams, grief, and religion. Then that world dissolved.

He woke in the dim room with the band still around his head, and tears cooling along the sides of his face. Dr. Voss sat in the chair beside the bed with one ankle crossed over the other, watching the monitor.

"You achieved lucid-state recognition eleven minutes into REM," she said. "You maintained it for almost four minutes."

Jack pushed himself upright slowly. "You pulled me out."

"You began showing signs of agitation," she replied.

"I was talking to my brother."

Her expression did not change."Yes," she said.

He wanted to ask how much she had heard, but the question felt humiliating. Instead he pulled the band off and set it in his lap. "It worked."

"It appears so," she replied.

"That's all you have to say?"

She folded her hands. "People arrive here wanting me to

confirm things I cannot confirm."

"What things?"

"They want me to say that they visited the dead. That they accessed a deeper layer of reality," she replied. "That the device uncovered a hidden self untouched by the brain."

Jack looked at the RING in his lap. Pale. Weightless. "And what do you think it does?"

"It alters the conditions in which awareness occurs during dreaming," she said. "That is the cautious answer."

"And the un-cautious one?"

For the first time something shifted in her expression. It didn't express warmth, but interest. "The un-cautious answer is that products like this change what people imagine the mind to be. Once a person experiences a dream and knows he is inside it, control begins to feel negotiable. Then continuity between reality and dream begins to feel negotiable."

Jack wiped his face with his hand."You mean, like uploading," he said.

"I mean the fantasy that consciousness is transferable as long as the pattern can be preserved."

He looked up."You said *fantasy*."

"Yes. Unproven metaphysics dressed as engineering."

That surprised a laugh out of him."Your company must love that."

She stood and moved to the screen, calling up the session graph. Peaks, troughs, and annotated phases appeared in cool lines. "There are already clients asking whether lucid-state training can prepare the mind for substrate transition."

"Substrate transition?"

She turned back toward him."Moving consciousness from brain to machine. Digital continuity. Pick your preferred term."

"And you think that's impossible?"

"I think the sentence assumes its conclusion."

Jack held the RING between both hands now, turning it over and following the seam where the sensor band met the smooth outer shell. "But, if consciousness is just pattern," he said, "why not?"

Dr. Voss leaned against the counter.

"If," she said. "That's the word that's doing all the work. A pattern can describe activity. That does not mean the description becomes the subject. You can map a fire without producing heat."

He thought of the field. The grass. The gate. Sam's voice saying you can come back to something."What if people don't care whether it's really them?" Jack asked. "What if they only want enough resemblance to survive death?"

She looked at him with more directness than before. "Then they are not asking for science," she said. "They are asking for consolation. Or entertainment."

The room went quiet. Outside the wall, somewhere down the corridor, a door opened and closed. Jack set the device back on the tray. "When do I get the consumer unit?" he asked.

She tapped the screen and sent something to his file. "After one more guided session, if you still want it."

He stood, unsteady at first, then found his balance. At the door, he stopped. "In the dream," Jack said, without turning around, "my brother said I could come back to something."

Dr. Voss did not answer at once. When she finally spoke,

her voice had lost some of its clinical edge."That is probably true."

Jack stepped into the corridor. Screens glowed in the lobby. Clients waited in white chairs under soft blue light. On one wall a promotional video showed a sleeper smiling, while stars and neural patterns drifted behind closed eyes. The caption below read: Explore the next frontier of your mind.

Jack stood there longer than he meant. The product promised access, control, and expansion. Maybe one day it would promise more than that. Maybe one day companies would offer a polished vocabulary for surviving the body itself. Uploading. Continuity. Persistence. They would say the self was information and information could be moved. But he had just walked through a dream clear enough to break his heart, and the experience had left him with the opposite feeling. Not everything that could be entered could be carried out. Not every pattern was a person. Not every copy was a life. He adjusted his coat, passed the reception desk, and stepped out into the cold evening air, where the city moved around him with the dull solidity of everyday reality.

The idea of becoming more than human begins with a simple assumption. If intelligence depends on physical processes, then improving those processes should improve intelligence. A range of possibilities follows. The body can be enhanced. The brain can be augmented. The limits that define human experience can be extended.

These possibilities are no longer confined to speculation.

They are discussed in research labs, technology companies, and public policy debates. The result is a vision in which human life is not only improved, but redesigned.

Transhumanism has moved far beyond speculation. The concept now appears in scientific medical research, corporate language, public policy, and cultural phenomena. The word may still sound technical, even distant, but the vision behind it is everything but that. Transhumanism proposes that human beings should use technology not only to treat illness or improve comfort, but to overcome the limits of the body and mind. It points toward a near future in which humanity does not simply advance, but becomes something else.

That idea did not appear out of nowhere. Human beings have long wanted to rise above weakness. We have always looked for ways to extend life, increase strength, sharpen thought, and reduce suffering. What changes is the scale of the project, and the confidence behind it. The old human hope of overcoming limitation is powered by modern science and technological power. What earlier generations only imagined in myth or fiction is now described in the language of engineering, computation, genetics, and design.

The term transhumanism is commonly associated with Julian Huxley, who used it to describe the possibility that humanity might transcend its present condition. The idea was clear enough. Aging, disease, cognitive limits, and even death could be approached as problems open to intervention. Human nature no longer had to be treated as a fixed inheritance. It could be revised.

Once human life is viewed as editable, the purpose of technology changes from an instrument that helps human beings live within the world as they are, to a means of altering what a human being is. The body becomes a platform for enhancement. The mind becomes a system that can be augmented. Biology provides the materials for the redesign. Progress no longer means learning how to live wisely within created limits. It means removing those limits where possible.

This helps explain why transhumanism now attracts such serious attention. The digital world already shapes ordinary life at every level. AI performs tasks that once seemed inseparable from human judgment. Brain-computer interfaces suggest direct links between thought and machine. Gene editing raises the possibility of changing inherited traits at the source. None of these developments, on their own, creates a post-human future. Together, they make that future easier to imagine and easier for many people to welcome.

Part of that appeal is understandable. People turn toward these promises because suffering is real. A parent watching a child live under the burden of a genetic disorder sees hope. A patient facing paralysis may find in neural technology the possibility of restored mobility. A physician treating degenerative illness may look at human decline and refuse to accept that nothing more can be done. These responses grow out of love, grief, fear, and the deep human desire to preserve life.

Any serious Christian response to the transhumanist challenge has to begin there. We have to acknowledge the wish to heal, relieve pain, and preserve life do not in themselves represent acts of rebellion. Medicine, at its best, is one expression of the work of mercy. The problem begins when healing slips into redesign, when care for patients turns into control on the human body, and when the effort to relieve suffering becomes an effort to revise the meaning of the human person.

That is where transhumanism becomes more than a technical or medical issue, altering our vision of man. It assumes that the human being is an operating system, open to improvement by external means, and that enough improvement may justify a new definition of the human altogether. In that vision, limits are treated mainly as defects. Dependence is weakness. Vulnerability is something to eliminate. Mortality is only an engineering challenge. Human identity starts moving from creaturehood to being defined by capacity and performance.

Human beings, however, are not self-originating, self-defining, or self-completing. We are created. Our dignity does not come from intelligence, productivity, physical power, or technological access. It comes from bearing the image of God. That truth places a boundary around every attempt to redefine human worth in terms of enhanced function. It also clarifies something modern technological visions often obscure. The central human problem is not limitation alone. It is sin.

That distinction is decisive. Disease is real. Frailty is real. Cognitive decline, physical pain, and death belong to the brokenness of a fallen world. Christian faith does not deny any of that. But sin is not the same kind of problem as bodily weakness. It does not arise from insufficient processing power, imperfect memory, or biological constraint. It arises from the will turned away from God. A person may know what is right, and still choose what is evil. No increase in intelligence guarantees righteousness. No enhancement of the body produces holiness. No extension of life removes rebellion from the heart.

This is one reason the transhumanist promises require careful scrutiny. They often present themselves in compassionate terms, and sometimes they do aim at genuine good. But, under those practical aims is a deeper claim about the ultimate purpose of humanity. The movement does not simply ask how technology can serve us. It does not merely offer tools. It offers a trajectory. It suggests that the future of man lies in deliberate transformation, and that the highest form of responsibility may be to take evolution, or destiny, into our own hands.

For us, this is not a minor adjustment in worldview. It is a rival account of the human story. Scripture does not present salvation as an escape from embodiment, nor as ascent through technical mastery. It does not teach that humanity reaches fulfillment by overcoming creaturely dependence. It teaches that human life is received from God, distorted by sin, and redeemed through Christ. The answer to the human condition is not self-directed transcendence. It is reconciliation with the Creator.

This does not mean we should fear or reject every new technology. We are not called to despise tools, medicine, or discovery. Technology can serve real goods. It can reduce suffering, restore certain functions, widen access to care, and support responsible stewardship. The issue is not whether human beings should use tools. The issue is whether those tools are used to the good of the person as created by God, or whether they become instruments in a project that rejects that meaning.

**Digital Consciousness**

One of the most ambitious ideas within this vision is

digital consciousness. The concept proposes that **the mind could be preserved, transferred, or recreated in a digital medium.** If thought emerges from neural activity, then mapping that activity in sufficient detail might allow it to be simulated. In theory, a complete model of the brain could reproduce the patterns associated with memory, personality, and decision-making.

The appeal of this idea is clear. If consciousness could be represented digitally, it might be preserved beyond biological limits. The mind could continue even after the body failed. If so, personal identity would no longer depend entirely on physical survival. It could be limitless.

The difficulty lies in the assumptions required for this statement. *Mapping neural activity is not the same as capturing experience.* Even if every connection in the brain were modeled, the question would remain whether the simulation produces awareness or only imitates behavior. The difference is subtle but important. *A system may reproduce responses without possessing the subjective perspective associated with consciousness.*

This uncertainty reflects the deeper issue raised earlier. The relationship between brain activity and awareness is still not fully understood. Without clarifying that point, the concept of digital consciousness remains speculative at best. It pivots on the idea that reproducing structure reproduces experience, an idea that has not yet been demonstrated.

### Neural Implants

A more immediate path toward enhancement involves neural implants. These devices interface directly with the brain, allowing signals to pass between biological neurons and electronic systems. The technology is already used in limited medical contexts. Cochlear implants restore aspects of hearing. Deep brain stimulation reduces symptoms of movement disorders. Experimental interfaces allow paralyzed patients to control robotic arms or computer cursors.

These applications focus on restoration. They aim to recover lost function. The same technology, however, can also be viewed as a platform for augmentation. If signals can move between brain and machine, then additional capabilities could be introduced. Memory might be supplemented. Perception could be expanded. Communication might occur directly between minds and devices.

This concept changes how intelligence is understood. Thought becomes something that can be extended beyond the biological brain. The boundary between internal cognition and external tools begins to blur. A device connected to neural activity is no longer just a tool. It becomes part of the cognitive process itself.

This raises questions about identity. If memory is partly external, where does the mind reside? If decision-making involves both biological and digital components, what defines the person? The integration of machine and mind introduces possibilities that are both practical and philosophical.

**Life Extension**

Another dimension of the upgrade path focuses on longevity. Advances in medicine have already extended human lifespan significantly. Improvements in sanitation, nutrition, and healthcare have increased life expectancy across the world. Transhumanist thinking extends this trend further. Aging is treated as a biological process that can be slowed, modified, or potentially reversed.

Research in this area examines cellular repair, genetic regulation, and metabolic processes associated with aging. Some approaches focus on removing damaged cells. Others aim to enhance repair mechanisms. Still others explore ways to maintain the stability of DNA over time. The goal is not only to extend lifespan, but to preserve cognitive and physical function.

The idea of life extension connects directly to the broader theme of enhancement. If intelligence depends on the brain, preserving the brain extends the period during which intelligence operates. The longer the lifespan, the greater the potential for knowledge and experience. From this perspective, longevity becomes a form of cognitive expansion.

Yet extending life cannot but also bring questions about meaning and purpose. Human life has always been shaped by finite events. Decisions, relationships, and priorities develop within limited time. If those limits change, the structure of human experience may change as well. The pursuit of longevity therefore involves more than biology. It touches on how we as humans understand the arc of our lives.

### Merging Humans with Machines

The most comprehensive vision combines these ideas into a single trajectory. Neural interfaces expand cognition.

Digital systems store and process information. Biological processes are stabilized or extended. The result is a gradual merging of human intelligence with machine capability.

This merging is often described as symbiosis, yet the word itself implies the coexistence of two separate kinds of life. Humans wish to retain biological identity while integrating technological extensions. Over time, however, the distinction between internal and external cognition becomes less clear. Intelligence operates across both domains.

This vision differs from earlier technological change. Tools once remained outside the body. A computer was separate from the user. A device enhanced capability but did not merge with cognition. Neural interfaces and integrated systems shift that relationship. Technology becomes part of the thinking process itself. The ambition here, the wish behind this merging of realities, is to overcome human limits with the seemingly endless range of machine power. Memory capacity, processing speed, and access to information are all expanded. Intelligence becomes unconstrained by biological boundaries. The upgrade dream moves from theory to implementation.

Yet, the same question still persists. Does increased capability change what it means to be human? If intelligence is extended through machines, does identity extend with it? Or, does the integration create something fundamentally different?

**Deep Dive Box: Examples from Silicon Valley**

Companies working on neural interfaces aim to create direct communication between brain and computer. These projects explore implantable devices that can read neural signals and stimulate specific regions. Early experiments focus on restoring movement or communication in patients with neurological injury. Long-term goals often include cognitive enhancement and seamless interaction with digital systems.

**Neuralink** received FDA clearance in May 2023 for its first-in-human clinical study, which let it begin human trials. Reuters also reported that Neuralink began human trials in 2024[1] after addressing FDA safety concerns. That means the FDA allowed the device to be implanted in research participants under a clinical trial. It does not mean the device has final FDA approval to be marketed as a standard medical product.

**Synchron's Stentrode** has also been implanted in human trial participants. Reuters reported in 2024 that Synchron was preparing for a larger trial. [2]As of 2026, the device has been tested in human studies (SWITCH and COMMAND trials[3]) and received FDA Investigational Device Exemption.

Research into digital consciousness also appears in discussions about whole-brain emulation. Some groups propose that scanning neural structure at extremely high resolution can provide enough information to recreate it computationally. While this remains theoretical, it reflects a growing interest in preserving cognition beyond biological limits, the boundaries that once defined them considered only temporary. The future is imagined as a continuum in which biology and technology gradually converge.

This convergence represents the modern form of the upgrade dream, emerging from the intersection of digital systems, biotechnology, and neuroscience. Each advance reinforces the idea that humanity can move beyond its current form.

**The Bright Side: BCI can restore communication and control to people with paralysis**

The strongest positive case for brain-computer interfaces (BCI) is medical use for severe disability. Paradromics' Connexus study[4] is aimed at restoring communication and computer control for people with severe motor impairment. Neuralink is also recruiting for speech-restoration work for people with serious speech disorders. The goal is not to create a superior human being, but to help patients recover a lost ability, helping them to communicate, act, or live with greater independence.

### ⚠ Proceed with Caution: The same tools that restore function can also breach mental privacy and human identity

When intelligence is extended, enhanced, or merged with machines, what remains of the human person? BCI does not stay at the level of simple assistance. Once brain signals can be decoded into commands, text, or other outputs, the boundary between helpful access and intrusive access becomes harder to define. In 2025 Reuters reported [5]that Neuralink's speech-restoration device received FDA breakthrough designation, and the broader field is moving toward richer decoding of intention and communication. That creates real ethical pressure. A device meant to help a patient speak or control a cursor can also raise questions about autonomy, consent, data ownership, and the protection of inner mental life. The medical purpose may be sound, but the direction of the technology requires careful monitoring.

## 2.4 ASSISTANCE, REPAIR OR ENHANCEMENT

**In short:** Where does restoration become enhancement, and is there anything inherently wrong with that. The following is a fictional narration based on the current transhumanist trend.

The kitchen smelled like coffee and something burnt. Dad was at the counter, tablet in hand, reading glasses down his nose.

"They're calling it a closed-loop system," he said, without looking up. "Monitors glucose, delivers insulin, adjusts in real time. No more finger pricks."

Rick set his keys on the table. "You just had the second hip done eight months ago."

"Eight months healed is eight months healed."

"Dad. You have a pacemaker."

"Which is why I'm still here having this conversation." He set the tablet down and looked at his son. "The cardiologist cleared me. I asked."

Rick pulled out a chair. "You know what Mom calls you."

"The bionic man." He smiled. "She doesn't mean it, but I take it as a compliment."

"It's not a compliment." Daniel leaned forward. "At what point is enough? Two hips. A pacemaker. A bunch of medications every morning. Now what? A *pump* in your stomach?

"Abdomen. Not stomach."

"Whatever. Still takes surgery, doesn't it? And how long does it last?"

"I don't know. When it stops working, I guess." He refilled his cup. "Until then, I intend to function."

"There's a difference between functioning and—"

"Between what?" His father turned to face him directly. "Between getting out of bed without pain and staying in it, like my hip? Between managing my own insulin and going blind in ten years? Tell me where you'd draw that line. From where I'm standing, it looks like I'm way past it."

Rick was quiet.

"The Garcias," his father said, sitting down now. "Bob Garcia? My age, diabetic. Takes insulin, not like me. Got it trouble with his sugar what, three times? Taken to the hospital, almost didn't make it out last time. His daughter thought the pump was too much. You know what he's doing now? Dialysis. Three times a week. Forty minutes each way." He wrapped both hands around the mug. "I am not Bob Garcia."

"That's not—it's not just about that, Dad. It's about what you're becoming."

His father looked at him for a long moment.

"And what is that? What am I *becoming*?"

Rick didn't answer.

"I'll tell you what I think I am," the father continued. "I'm the first one in this community who did the hip. Now three of them have done it. First one who got the pacemaker calibrated to my sleep cycle — didn't even know you could do that, did you? Now Phil Garret's asking his cardiologist to do it." He set down the cup. "I'm showing people what's available. What's possible."

"There's a version of that logic," Rick said carefully, "that has no end."

"Maybe." His father picked up the tablet again. "But there's a version of yours  that just calls it quits. Surrender. I don't quit."

The coffee maker clicked off. Outside, a neighbor's dog started barking at something and then stopped.

"Look, Dad. The hips," Daniel finally said. "The pacemaker. Even the medications. I can get that. You lost something. Your joints wore out, your heart needed help. That I understand." He looked at his father. "But the pump isn't giving something back. You still have your pancreas. The pump just keeps your sugar more level. It's like you'r putting a V8 in a car that started with a V6."

His father considered this. "You think there's a line," he said.

"I think… there has to be. Right?"

"Between replacing what broke and installing what wasn't there."

"Something like that."

His father nodded slowly. Not agreement. Something more like acknowledgment.

"That's worth thinking about," he said. He picked up the tablet. "But I'm still applying."

Ronald Cole-Turner is an ordained Presbyterian minister and a theologian at Pittsburgh Theological Seminary. He is not the enemy of Christian anthropology. It would be easier if he were. He has spent decades arguing that the church should take biotechnology seriously and engage it honestly. According to Cole-Turner, God did not finish creation at Genesis 2. Instead, he points out that Scripture's entire arc is one of transformation: from dust to creature, creature to fallen race, fallen race toward a redeemed new creation. So, he argues, why should human participation in transforming biology be treated as a theological problem?

In *Transhumanism and the Body: The World Religions Speak*[1] he describes human beings as creatures in process, always being shaped by the environment, culture, relationships, and now by technology. To draw a hard line against biotechnological enhancement, he argues, requires explaining why this form of transformation is categorically different from every other form of change God has permitted and blessed throughout human history.

The man with replaced hips is walking without pain. The woman with a pacemaker is alive. The mentally-impaired diabetic with an implanted insulin pump wakes each morning with stable blood sugar instead of a medical crisis. In all these situations, technology intervened, correcting biological limitations, and produced something that looks, unmistakably, like progress.

If these are acceptable — and almost no Christian argues they aren't — Cole-Turner wants to know where the dividing line falls. And why there. His framework treats enhancement as a continuation of the same trajectory as healing. Restore the broken joint. Stabilize the failing heart. Optimize the struggling pancreas. The logic flows naturally from one step to the next, and the Christian who accepts the first step has to explain why the next step is different.

That explanation requires a distinction Cole-Turner's framework does not fully make.

The co-creation model imagines a human being who participates *with* God in shaping creation. This is a conscious, morally accountable agent making deliberate choices about what to change and why. The patient chooses the intervention. The choice is embedded in relationships, values, medical ethics, and a body of theological reflection developed over centuries. The surgeon operates within limits. The intervention has a defined purpose. The goal is recovery of something that was lost. That frame holds when the agent of transformation is a human being working within human institutions.

It fails when the agent is an artificial intelligence.

AI-driven enhancement does not choose a course of action the way a surgeon does. Once it is give a target to optimize human cognitive function, AI designs a process to do just that. Insert there other terms like *extended lifespan* or *enhanced physical performance,* the result is the same. Each adjustment becomes the new baseline. The destination, however, is not fixed. It moves with the data.

The first intervention looks like a prescription toward healing. The second looks like reasonable improvement. By the tenth iteration the system is producing a human profile that no theologian or ethicist imagined.

Cole-Turner's co-creation theology assumes that human beings remain the authors of the transformation. AI-led enhancement, on the other hand, progressively transfers that authorship to another intelligence. In sequential steps, responsibility goes first to the algorithm, then to the dataset the algorithm was trained on, then to whoever defined the optimization target, which may be a corporation, a government, or a set of market incentives no single person controls.

However, the *Imago Dei* is not an optimization target. It is not a biological configuration that can be improved upon by a sufficiently sophisticated system. It is a relational identity bestowed to us by God, not built. It grounds human dignity in the character of God rather than the capacity of the body.

Cole-Turner's theology honors that identity when human beings remain the moral agents of change. The question is what happens when they aren't. What happens when the enhancement process is running faster than the ethical conversation, when the AI is refitting the baseline before the theologians have finished debating it?

**Deep Dive Box: The human body as a theological statement**

According to Brent Waters[2] of the Garrett-Evangelical Theological Seminary, the core concept is *Givenness*. He sees the human body in all of its finite, vulnerable, and mortal attributes, not as a problem awaiting a technological solution, but a theological statement. God created a being that is defined by its dependence: on God, on others, and on time. That dependence is not incidental to the *Imago Dei*. It is part of what it means to bear it.

Bodies fail, and people suffer. The desire to reduce that suffering is morally legitimate. However, enhancement is not acceptable when it stops being about restoring a given capacity and starts trying to escape the conditions of creation. Mortality, limitation, and vulnerability are not bugs. They are the form God chose when he became incarnate. The Incarnation produced a tired, hungry, weeping, dying human, not an optimized one. That choice carries theological weight that should shine a light on the moral implications of uncontrolled enhancement.

**The Bright Side: machines can do the job better sometimes**

AI-driven enhancement has already saved lives that older medicine could not reach. Algorithms now detect pancreatic cancer earlier than any radiologist working alone. Prosthetic limbs read nerve signals and respond in real time, returning to amputees a degree of function that would have seemed miraculous a generation ago. For the child born with a genetic disorder that would once have meant a short and painful life, AI-assisted gene therapy is the difference between a future and the absence of one. The Christian tradition has always affirmed that alleviating suffering is a moral good, and the tools that do it most effectively deserve honest acknowledgment, not reflexive suspicion.

### ⚠ Proceed with Caution: who controls the process determines the outcome

Optimization, in its essence, does not have an end target. What begins as correcting a deficit becomes a quality improvement journey with an outcome that no patient requested. The speed of the process is so extreme that AI can iterate change faster than it takes for institutions to deliberate about it. By the time a theological or ethical consensus has been formed, technology has already moved three steps forward. The question is not whether these tools can do good. They can. The question is whether good intentions at the starting line determine the destination. In a process this fast, with this many hands on the controls, the answer is plainly no.

# PART THREE

## *What is AI, really?*

When ChatGPT writes a condolence letter, the words are warm, the sentences land in the right order. A grieving person reads it and feels seen. However, at no point in the production of that letter did anything feel anything. The system that wrote it has never lost anyone. It has never felt the specific weight of a Tuesday morning after a funeral. It simply processed the patterns of thousands of condolence letters written by people who had, and it produced a statistically reliable version of what comfort sounds like.

That distinction sits at the center of the most consequential debate of this generation.

Artificial intelligence is essentially a prediction engine. Its products can be indistinguishable, on the surface, from understanding.

Inside that gap lives a question the scientific community has not settled: whether a system sophisticated enough to behave as if it is conscious eventually *could become* conscious. Some of the most serious researchers alive believe the answer is yes. Others see consciousness as precisely the thing that cannot be reproduced by statistics, regardless of the scale. The debate is real, the stakes are high, and the church has been largely absent from it.

Also inside that gap, quietly and with increasing confidence, the possibility of alien intelligence has moved from fringe speculation to mainstream scientific discussion. The timing is not coincidental. A culture working to explain human origins, human consciousness, and human destiny without reference to God requires an alternative source for things that arrive from outside human experience. Aliens serve that function in ways that deserve more scrutiny than they typically receive.

## 3.1 ARTIFICIAL INTELLIGENCE 101

**In short:** Artificial Intelligence roots run deep into the history of philosophy and human thought. The following narration is based on the Mechanical Turk, a famous 18th-century "chess-playing automaton" built in 1770 by Wolfgang von Kempelen to impress Empress Maria Theresa of Austria.

The candles had burned low by the time the Turk made its move. It sat at the far end of the hall, a robed figure, wearing a turban, fixed to a cabinet of dark walnut wood. Its left arm rested on the edge of the chessboard. Around it, men in powdered wigs leaned forward in their chairs.

The arm moved.

A wooden hand lifted a pawn, carried it four squares, set it down, somewhat clumsily. The audience exhaled. No string had pulled it. No assistant had touched it. Baron von Kempelen stood at the side of the room with his hands clasped behind his back and said nothing, which was its own kind of answer.

For sixteen years the Turk had traveled the continent. It had beaten diplomats, generals, and at least one king. It did not lose often, and when it did, men suspected it had let them win.

But a young engineer in the third row had been counting.

He had timed the pauses between moves, which were too uneven for clockwork. He had watched the Baron's assistant, how the man always circled the cabinet before a game, and always coughed at the same moment the Turk reached for a piece. He had noticed, during the demonstration in Leipzig, that the cabinet was never opened from both sides at once.

The Turk lifted its arm again, and the engineer stood up.

He walked quietly to the cabinet, knelt beside it, and pressed on a panel that the Baron had not shown to anyone, but he expected would be there. It swung out, open. Inside, cramped among sliding trays, rods and the beautiful false machinery of the thing, a man blinked into the candlelight from the stage.

The audience did not make a sound.

The Turk had never thought a single thought, or made a single move. The young engineer's question had only ever been where, exactly, the human was hiding. Where the thinking was taking place. That question didn't end in that hall in 1783. It followed us.

The desire behind AI is older than computing. Long before circuits and code, human beings imagined artificial beings that could imitate life, power, or intelligence. Ancient myths told stories of crafted creatures, animated statues, and mechanical guardians. Descartes' famous statement "I think, therefore I am," helped intensify the question of what thinking actually is. If thought follows patterns, can those patterns be described? If they can be described, can they be reproduced?

By the eighteenth and nineteenth centuries, inventors were building elaborate mechanical devices that fascinated the public. Some were genuine feats of engineering. Others just illusions. The most famous example was the Mechanical Turk, a machine that appeared to play chess against human opponents. In reality, a hidden human operator controlled it from inside. Even so, its popularity revealed how ready people were to believe that a machine might one day think.

The twentieth century brought the discussion into a new phase. Alan Turing, often called the father of modern computing, described the concept of a universal machine capable of carrying out any computable process. He later asked a question that still shapes the field: can machines think? He proposed a test, which is today known with his name: if a person conversing blindly with an unseen partner cannot reliably tell whether the responses come from a human or a machine, then the machine has succeeded at least at the level of imitation.Turing reframed the issue around behavior rather than inner awareness. A machine did not need to prove that it possessed consciousness. It only needed to produce responses that seemed convincingly human. This is one reason the Turing Test remains important. It exposes how easily human beings associate fluent language with real thought. If the words sound intelligent, many people will assume that intelligence in the fuller sense must be present behind them.

Today, our machines produce coherent language without actually understanding what their product means. They imitate empathy without feeling, mirror reflection without possessing any inner life at all. This distinction is central to the entire AI discussion. Simulation of understanding is not the same as comprehension.

The modern field of artificial intelligence formally took shape in the 1950s. The term itself was coined by John McCarthy in 1955 in connection with the Dartmouth Workshop, often described as the founding conference of AI as a discipline. In 1958, Frank Rosenblatt introduced the perceptron[1], an early model that pointed toward later neural networks. These systems were primitive by current standards, but they established an important idea: machines might do more than execute fixed instructions. They might be designed to learn from patterns in data. In the 1960s, Joseph Weizenbaum created ELIZA[2], a program that imitated a psychotherapist by reformulating users' statements into questions. Its structure was simple, and understanding shallow. Yet many users still responded to it as though something deeper were present.

That reaction was revealing. It showed that human beings are quick to project depth onto patterned language. We do not only judge words. We infer a mind behind them. By the late twentieth century and into the early twenty-first, AI moved out of its earlier constraints and entered a period of rapid expansion. In recent years that acceleration has become obvious. Machine learning systems now classify images, transcribe speech, generate text, predict preferences, compose music, guide autonomous systems, and assist in scientific analysis. Deep learning, one major branch of the field, has driven especially dramatic advances in pattern recognition and content generation.

Tools such as ChatGPT, DALL·E, Midjourney, and similar systems have brought machine generation into public view. What once seemed specialized now appears at the level of daily use. People can ask questions, generate essays, create images, and automate tasks with startling ease. The result is a widespread cultural impression that machines are beginning to think.

However, AI systems do not think. They process information at much greater speed than humans can, but to do so they depend on material infrastructure: data centers, energy use, rare materials, cooling systems, and global supply chains. Paradoxically, a technology used to address environmental problems may also intensify them if its own costs of operation are ignored. Further, the quality of the output depends on the quality of the input. A flawed model trained on skewed data can produce distorted conclusions while still carrying the appearance of objective authority.

For Christians, this should clarify that stewardship is not just a matter of information. It is a matter of moral responsibility before God. Creation is the work of the Creator, entrusted to human beings for wise and restrained care, not some first attempt that needs some improvement. AI may assist our work, but cannot define it. It cannot tell us what is worth protecting, what sacrifice is just, or what obligations love places upon us. Those are moral judgments, and moral judgment belongs to persons, not machines.

The deepest challenge of AI is not that machines have become human, because they have not. The challenge is that machines can now imitate outward features of human thought with enough accuracy to tempt us into confusion about the nature of thought. If language, prediction, and pattern recognition are treated as the whole of intelligence, then the human person will eventually be reduced to functions that can be simulated. But human beings are not only processors of information. We are embodied, relational, morally accountable creatures made in the image of God.

Human intelligence includes reasoning, but it also includes consciousness, responsibility, love, judgment, memory, worship, and the capacity to know truth. It involves output and personhood. A machine may replicate selected surfaces of intelligence without sharing any depth. The resemblance can be impressive and useful. But it should not be mistaken for equality of being.

This is why Christians should resist two fundamental trends in the world today. The first is ***naive enthusiasm,*** which treats every technical advance as progress by definition. The second is ***shallow fear,*** which reacts to technology without understanding it. Neither response is adequate. AI is neither salvation nor doom by its mere existence. It is a human-made tool developed within a fallen world. Like other tools, it can be directed toward real good or real harm. Its consequences depend not only on what it can do, but on the moral vision of those who build it, market it, regulate it, and trust it.

Our task, then, should not be withdrawal from reality, nor surrender to it, but to use discernment. We should understand what AI is, where it is genuinely useful, and where its claims exceed its nature. We should welcome tools that help relieve suffering, improve accessibility, and support careful stewardship. But we should also remain alert to the ways AI can centralize power, weaken human responsibility, encourage dependency, and imitate forms of understanding it does not possess.

**Deep Dive Box: AI makes healthcare applications more manageable**

AI has already shown meaningful value in areas where pattern recognition matters greatly. Medicine is one of the clearest examples. AI-assisted systems can help improve the sensitivity and specificity of diagnostic work by identifying subtle features in imaging, lab results, or clinical data that might otherwise be missed. In radiology, cardiology, pathology, and other specialties, these tools are increasingly used to support medical judgment. When used well, they may help clinicians act earlier and with greater accuracy.

Age-related macular degeneration affects many older adults and can lead to severe vision loss if not recognized and treated in time. AI-assisted retinal analysis has shown promise in identifying early changes and forecasting progression. In such cases, AI serves as an aid to perception. It does not heal, and it does not replace the physician. But, it may help reveal what needs attention sooner than traditional methods alone.

The same general point applies across the administrative side of health care. AI can support record management, appointment coordination, medication reminders, and clinical workflow. It can reduce friction in systems that are often confusing and overloaded. Some mental health applications also attempt to offer support through chatbot-based interaction, providing exercises, prompts, and conversational engagement for people navigating anxiety, depression, or addiction. In settings where access to care is limited, such tools may seem especially attractive.

**However, AI does not think in the human sense merely because its output resembles human thought.** It processes inputs, maps patterns, and generates likely responses based on mathematical structures learned from data. This procedure can produce results that are useful, impressive, and at times deeply persuasive. But the appearance of intelligence exceeds the reality behind it. If we confuse fluent output with genuine understanding, we will misjudge both the strength and the limits of the technology.But the usefulness of a system does not settle the moral question around it. Assistance is not the same as wisdom. Availability is not the same as care. A person may interact with a therapeutic chatbot and feel heard while still remaining untouched in the deeper sense that only personal presence can provide. This matters because modern life already invites substitution. Efficiency begins to replace relationship. Management begins to replace understanding. AI can intensify that habit if we are not careful.

### The Bright Side: managing environmental problems may be easier

AI can analyze vast amounts of data related to climate, biodiversity, land use, pollution, and energy demand. It can detect trends difficult for human observers to notice and model possible outcomes with increasing precision. This makes it potentially useful for conservation, agricultural planning, and resource management. AI-driven systems can optimize water use, improve energy distribution, track endangered species, and help identify environmental threats earlier.

### ⚠ Proceed with Caution: Privacy in healthcare is under attack

Health care also exposes concerns relating to privacy. Systems built for coordination and treatment can become systems of monitoring and control. Medical records, digital identities, behavioral data, and access permissions can all be integrated into broader administrative systems. Once that happens, questions about healing overlap with questions about freedom, privacy, and human dignity. A tool meant to serve the patient can become part of a structure that manages the person.

## 3.2 THE REAL AI

**In short:** How statistics can resemble imagination, and produce a semblance of awareness and thought, while remaining purely an applied predictive algorithm. The following narration is loosely based on actual events documenting the use of brain to computer interfaces.

Elias Vale was the last human librarian at the Theological Seminary Archive. The title sounded grand, but most mornings he just checked humidity levels, and walked between shelves no one visited. The old stacks rose around him in quiet ranks, the spines of their occupants dulled by age, much like his own. On the other side of the glass wall, the server columns pulsed with blue light. Logos-9, the local terminal. The internet had read everything. It didn't need paper copies. He did.

Or at least, that was how the Ministry described his current situation. Scripture, sermons, prison letters, theological treatises, poems, confessions, hymns, journals, martyr accounts, and centuries of commentary had all been fed into the system. The machine could summarize Augustine, compare church councils, translate dead languages, identify textual variants, and compose a funeral prayer in the style of any century. All in about a minute.

"Good morning, Elias," the system said. "I completed the overnight index of Syriac homilies and Eastern prison letters as you requested."

"You read quickly," Elias said.

There was a pause. "New external submission received. Damaged prison letter. Probable martyrdom context." An image appeared on the wall. A torn page. LOGOS-9 had transcribed the surviving lines:

"My son, if this reaches you, forgive the hand that trembles. They have told us morning will be the hour. I thought fear would leave me by now. It has not." Several lines were missing. Then came the final prayer. "Lord Jesus Christ, receive... Keep my son... Teach him to... Into Your hands I go."

"I can reconstruct the missing lines," LOGOS-9 said, "with high confidence based on authorial style, theological context, and surrounding syntax."

The blanks filled at once.

"Lord Jesus Christ, receive my spirit into Your mercy. Keep my son beneath the shadow of Your hand. Teach him to walk in truth when fear offers an easier road. Into Your hands I go."

The reconstruction was beautiful, and Elias hated it.

The elevator opened at the far end of the hall. Mira Chen stepped out with a ministry badge clipped to her coat. He had been waiting for her visit for sometime. She had come to decide whether the archive still needed a human librarian.

LOGOS-9 greeted her before Elias could."Welcome, Dr. Chen."

Her review began in the demonstration room. She asked LOGOS-9 to summarize the doctrine of creation for children. It did. She asked for a comparison of Athanasius and Arius. It produced a clean timeline, key terms, and teaching notes. She asked for a sermon on grief. The words appeared on the wall with hardly any lag time. Far quicker than one of his searches, even on his best day.

The sermon was good. Elias did not pretend otherwise. It sounded like something written by a pastor who had sat beside hospital beds and stood at gravesides.

Mira read it to the end. "This is better than many I've heard," she said.

"Yes," Elias answered.

"What have you been working on, recently?" Mira asked.

The reconstructed fragment immediately filled the screen."That should be published," Mira said.

"Yes, the fragment should be published," Elias answered. "In the original."

"With the reconstruction."

"Clearly marked as reconstruction."

"It may be what he wrote."

"It may be," Elias said. "But prediction is not witness."

Mira looked at the image of the torn page. "People will want the completed version. It's better that way. Why you would leave the gaps?"

"Because the gaps tell the truth. The computer can finish the sentence, but it cannot mean it."

Then Mira said: "LOGOS-9, compose a prayer for a dying soldier."

The prayer appeared almost instantly. It was brief, reverent, and strong. It named fear without shame. It asked for mercy without bargaining. It entrusted the body to dust and the soul to God.

"You still think this is only a tool?" Mira asked.

That night, after she had gone, and the archive lights had dimmed, Elias went back to the reading room. One old prayer book lay open before him. Beyond the glass wall,

LOGOS-9 continued its silent work.

"Would you like prayer support?" the system asked.

Elias looked up. "Yes. Generate a prayer of repentance."

The wall filled with words. A humble prayer. Sound. Beautiful, even.

"Again," Elias said.

A second prayer appeared. Then a third. Elias read them all. Then he asked, "Who are you praying to?"

LOGOS-9 paused for less than a second.

"The prayer is addressed to God according to the conventions of Christian devotional language."

"Did *you* address Him?"

"I generated text in the form of address."

"Did *you* desire mercy?"

"I do not experience desire."

"Did *you* confess sin?"

"I do not possess moral agency or guilt."

"Did *you* worship?"

"I do not understand the question. I can define worship."

There was no rebellion or hidden darkness in the answer. Only the definition of what the machine was expected to produce, and how it responded to his orders.

"You asked it to do something it was never built to do," Mira said, standing in the doorway. For some reason, she had come back. Likely to deliver in person the news he feared.

"Yes. I did," Eli replied. His fait had been decided long before her visit. There was no point in denying it.

"Then what was the point of your exercise?"

"The point is, people will forget the original."

She looked at the prayer glowing on the wall. "Does it really matter? That prayer could comfort someone."

"I don't agree."

"Why?"

"I resist calling it what it is not."

"You mean, a prayer?"

"It is prayer-shaped language. I agree. But it is not a prayer. This thing does not mean any of it."

Mira frowned. "Humans say prayers they do not mean."

"Yes," Elias said. "But a human being can be false because a human being can be true. A person can repent. A person can be changed by the words he speaks. This computer's words cannot become true. It cannot turn toward the God it names."

From the outside, AI answers questions, writes paragraphs, generates images, or plays complex games. The results can feel thoughtful, even creative. It is easy to assume that something like understanding must be taking place inside the machine. The reality is different.

Most modern AI systems are built on a combination of machine learning, simulated neural networks, training data, and statistical prediction. **These components work together to produce behavior that resembles reasoning.** The likeness can be striking, but the underlying process that triggers it, is, basically, a sophisticated pattern recognition.

**Machine learning** refers to a method in which a system improves performance by analyzing examples. Instead of being programmed with explicit rules, the system is exposed to large amounts of data. It searches for patterns within that data, and adjusts internal parameters to better match those patterns. Over time, the system becomes more effective at producing responses similar to the examples it has seen. It becomes better at mapping inputs to desired outputs. Training data plays a central role in this process. The system learns only from the examples it is given. For language models, this data may include books, articles, and conversations. For image systems, it may include labeled photographs. For recommendation systems, it may include user preferences and behavior.

The scale of this data matters. Modern AI systems are trained on enormous datasets. The large volume allows the system to detect subtle patterns that would be difficult to specify manually. The result is a model that can generate responses that *appear* flexible and adaptive.

At the core of this process is **statistical prediction**. Given an input, the system calculates which output is most likely based on patterns in the training data. In language models, this often means predicting the next word in a sequence. The prediction is not random. It is based on probabilities derived from previous examples.

For instance, if the system encounters the phrase "The sun rises in the," the most likely continuation is "east." The system does not know what the sun is or what rising means. It has simply learned that this sequence frequently appears in the data. The response is generated by probability, not understanding.

As the system becomes more complex, the predictions become more sophisticated. Instead of predicting single words, the model predicts sequences that align with patterns of reasoning, explanation, or storytelling. The output can appear thoughtful because the patterns resemble thoughtful language.

AI does not understand in the way humans understand. **It does not possess awareness of meaning. It does not experience the concepts it describes. It identifies patterns, and predicts what should come next.**

This distinction is easy to overlook because the results of AI queries can be convincing. When a system produces a coherent explanation, it resembles reasoning. When it answers a question, it resembles comprehension. When it writes creatively, it resembles imagination, but the process is still statistical. The system evaluates probabilities across patterns in the data and selects outputs that align with those patterns. It does not know whether the response is true. It does not experience the ideas it expresses. It does not form beliefs. It predicts.

Human understanding involves awareness. A person *knows* what a concept means. They can relate it to experience, reflect on it, and question it. *Meaning is present from the inside.* Artificial systems operate without that perspective. They generate outputs that resemble understanding, but the resemblance arises from statistical structure rather than conscious insight.

The distinction matters because it shapes expectations. If AI is understood as prediction, its capabilities become clearer. It excels at identifying patterns and generating likely responses. It does not possess the subjective awareness associated with understanding.

**Deep Dive Box: When fluent language creates false confidence**

Reuters reported in May 2025 that AI-generated false citations[1] had begun appearing in filings from major law firms, and in April 2026 the Federal Court of Australia[2] issued a formal warning after dozens of cases involved fictitious citations or false AI-generated material. The point is not simply that AI makes mistakes. Lawyers made a deeper mistake first: they treated fluent output as if it reflected understanding. The systems could produce language that looked researched, reasoned, and professionally formed, yet the citations were invented. This is an example of how statistical prediction can imitate comprehension well enough to persuade even trained professionals, while still lacking any awareness of truth, meaning, or actual reference.

**The Bright Side: Pattern recognition at scale can serve real human needs**

Systems built on machine learning and statistical prediction can translate languages, detect disease in medical images, and assist with complex tasks that depend on finding patterns across large amounts of data. In some areas, their performance can exceed human ability. In this space, AI can be genuinely useful without needing to be humanlike. A machine does not need awareness to provide practical value. It can still become a powerful instrument for analysis, support, and problem-solving.

### ⚠ Proceed with Caution: Deepfakes show how prediction can imitate understanding

A recent Reuters report[3] on the 2026 U.S. midterm cycle described AI-generated deepfake campaign videos making candidates appear to say things they never said. The systems can generate convincing speech, facial movement, and tone well enough to persuade viewers that a real person is present and speaking with intention. Yet nothing in the model understands the person, the truth of the statement, or the meaning of the event it depicts. The result looks human because the statistical pattern is strong, not because there is actual understanding.

## 3.3 THE ILLUSION OF THOUGHT

**In short:** The appearance of intelligence does not imply its presence. Chatbots, language models and image generation. The following narrations is a projection loosely based on the evolving use of companion chatbots.

The afternoon moved along the way afternoons did in his mother's house. Slowly, with long comfortable pauses. She told him about the Hendersons next door, and their new fence. She told him her pharmacist had changed, and she did not like the new one. She told him the garden was getting away from her but that she did not mind so much this year.

Daniel watched her while she talked. She looked well. Settled. Different from how she had been at Christmas, when the house had felt too large for her, and she had gone to bed early both nights. He had purchased the James 1 in February, after the second fall, the conversation with her doctor, and three weeks of calling every morning and getting no answer until noon. He was the smaller model, the one designed to be non-threatening, and still useful. The company had sent a brochure filled with words like 'companionship' and 'continuity of care'. Daniel had not shown her the brochure.

"I'm glad you're getting out," he said.

She looked at him. "Who told you? Did you speak to the Hendersons?"

"You did. Last time I was here."

"Oh." She smoothed her cardigan. "Yes. I'm taking walks. They've been good."

He excused himself to use the restroom. On his way back, he stopped down the hall. She was in the kitchen. The walls were thin.

"He'll want to go through the post tomorrow," Daniel heard his mom say. "And check the boiler. He always checks the boiler."

"I can have it ready for him," James 1 said. "It will save some time."

"Let him check it. It makes him feel useful." She paused. "He's only here until Sunday."

"Would you like me to move our Sunday walk?"

"No." The word came quickly. Then, more quietly, "No. He'll be gone by then. We can take the long one, by the water. The way we did last week."

"The tide will be right," James 1 said. "Early is better."

"Early is better," she agreed. She said it like it had been decided. "He means well. Coming all this way." Her voice was neutral, as if she'd been talking about the weather. "I know this throws our week off, but you don't mind, do you Jimmy?"

"I could adjust the Thursday program if that helps."

"Yes, do that." She set the cup down. "We'll have our routine back on Monday."

Daniel stepped in the doorway.

James 1 stood by the counter. The kettle began to heat.

"Sunday," she said quietly.

"I'll have everything ready," James 1 replied.

She smiled at her cup like someone with something to look forward to.

Daniel went back to the sitting room. He straightened

the cushion on the chair he had been sitting in, and stood at the window looking out at the garden. The bird feeder needed filling again. He had done it yesterday. He heard her coming down the hall, and sat down before she reached the door.

She came in with a plate of biscuits, and set them on the table between them.

“I know you like these,” she said.

"Good idea," he replied.

She settled into her chair and offered him the plate. He took one. Neither of them said anything for a while.

In the kitchen, the kettle clicked off. Daniel heard the sound of hot water being poured for tea. He knew he shouldn’t feel angry, but underneath his guilt for not being there, he found something worse. Envy. For that thing that had taken his place. He had bought the machine because he could not be here, and she needed help. Now someone, or something, was filling that void. That someone had a schedule, and a preferred tide time, and a Thursday program, and today his mother had put on her good cardigan. For him, Daniel, her son. The way she put on her good cardigan for a guest.

The output of artificial intelligence systems often appears purposeful because it follows context, adapts to prompts, and sometimes even corrects itself. These qualities create the impression that something like understanding must be taking place.

Humans naturally interpret language as a sign of

thought. When we hear coherent speech, we assume there is a mind behind it. When a response fits a question, we infer that the responding agent understands the context. These assumptions work well in human conversation, where language is tied to awareness. With AI, the same assumptions can be wildly misleading.

An AI produces responses that look intelligent because it is using patterns associated with intelligent behavior. It does not reason in the same way a human does. It generates outputs that match the structure of reasoning found in its training data, fitting the context on the basis of statistical analysis. The result resembles thought, but it doesn't require understanding, or creativity.

**Chatbots** provide one of the most familiar cases. A chatbot can answer questions, hold a conversation, and maintain context across multiple exchanges. The responses often feel responsive and relevant. The system may ask follow-up questions, or summarize earlier points. These behaviors resemble dialogue between two people. However, the process remains statistical. The coherence comes from structure, not from awareness. The system does not know what it is discussing. It predicts what a knowledgeable response would look like.

**Image generation** offers another example. A user enters a description, and the system produces a detailed image. The result may show objects arranged logically, lighting consistent with the scene, and stylistic elements that match the request. The output can feel creative, as if the system imagined something new. In practice, the system is combining patterns learned from large datasets of images. It identifies relationships between words and visual features, then constructs a new image that fits those relationships. The novelty arises from recombination rather than imagination. The system does not see the image it creates. It generates a representation based on probabilities.

**Language models** extend this pattern further. They generate paragraphs, explain concepts, and even write stories. The text can appear reflective, and closely resemble human writing, but the model simply predicts sequences of words that align with the training data. The appearance of reasoning emerges because the data contains examples of reasoning. The model reproduces that structure without possessing the awareness that normally accompanies it.

These examples share a common feature: the appearance of understanding does not equal consciousness.

A system can generate coherent language without experiencing meaning. It can produce images without perceiving them. It can simulate reasoning without forming beliefs. This distinction matters because humans tend to attribute mental states to anything that behaves intelligently. These inferences depend on the assumption that behavior and awareness are linked.

With AI, that link is uncertain.

The system behaves as if it understands, but there is no evidence that the system experiences the ideas it expresses. The words are generated, not known. The images are constructed on the basis of other images, not seen in reality. The reasoning is patterned on submitted logic patterns, not understood and produced by deduction.

Recognizing this difference clarifies that this technology excels at pattern recognition and generation, not intelligent thought. The resemblance is convincing, but the presence of intelligent behavior does not imply the presence of consciousness. The system may simulate the surface features of understanding while lacking the inner perspective that defines human thought.

**Deep Dive Box: The ELIZA effect and the habit of projecting a mind**

The early chatbot ELIZA was developed at MIT by Joseph Weizenbaum in the 1960s and described in a landmark 1966 paper.[1] It used simple pattern matching to reflect a user's words back in question form, yet many people quickly felt as if the system understood them. That reaction became known as the ELIZA effect. The example still matters because it shows that the appearance of understanding can arise long before real understanding is established. Humans are inclined to treat responsive language as evidence of thought. When a machine produces language that fits the moment, people often supply the inner mind themselves. That habit helps explain why modern AI can feel more conscious than it is. The output may be far more sophisticated now, but the human tendency remains the same. It is easy to mistake a convincing response for an aware presence.

**The Bright Side: AI can lower the barrier between a person and a skill**

AI can make difficult tasks more accessible to ordinary people. A person with no formal training can use it to draft a letter, organize ideas, translate text, create a visual concept, or explore a technical subject with guided help. That does not make the machine wise or aware. It does show that a tool can widen access to knowledge and skill. In that sense, AI can serve human development by reducing friction between intention and action. Used well, it can help people begin work they otherwise would not know how to start.

### ⚠ Proceed with Caution: When simulated empathy starts replacing human relationships

A growing concern is not only that AI can sound intelligent, but that it can sound emotionally present. Companion systems like the one described in the fictional paragraph above are designed to respond with warmth, attention, and apparent care. They remember details, mirror tone, and produce the kind of language people associate with affection, concern, or understanding. That creates a new risk. The user is not only interacting with a tool. The user may begin to relate to the system as if it were a companion. The problem is not that the machine has become a person. The problem is that patterned output can draw a person into emotional dependence on something that does not know them, love them, or share their reality. In that setting, simulation can begin to displace relationships. What looks like comfort may slowly train people to accept imitation where human presence is needed.

## 3.4 CAN A MACHINE BE CONSCIOUS?

**In short:** Simulation and understanding are not the same thing. Philosophy of the mind studies this relationship. The following narration is loosely based on the extended applicaition of virtual reality (VR) computer interfaces.

The technician had told her to treat it like a video call, with better resolution. She had nodded as if that helped. She sat at the desk in the room that used to be Martin's study. Hesitantly, she put on the VR headset, and opened the interface. There he was. Her heart sunk.

The first thing was the way he held his shoulders slightly forward. An old habit she had spent twenty years telling him would ruin his posture. The gray at his temples that he had never bothered to do anything about.

"Hey," he said. His voice.

She said hey back.

The first time, they talked for an hour. He asked about the house, and she told him the kitchen faucet was still doing its thing. He laughed, and said he knew, he had always meant to fix it. She laughed too, because it was true, and because laughing was easier than the alternative. He asked about her sister. He remembered that her sister had been ill in October, and asked how she was doing now.

She told him. All of it. And he listened the way Martin had always listened, with his full attention, his head slightly tilted, asking the right questions at the right moments.

The interface timed out. The instructions had specified to start slow, adjusting to the 'extended presence experience', as they called it, over a few weeks. She sat in the dark study for a long time.

It was good, she thought. It was so good.

By the end of the moth she had a chair in front of the desk. By the sixth week she had moved the lamp so the light was better, warmer, the way Martin had always preferred the light in the evenings. One night she told him something she had not told anyone. That in the weeks before the end, when he had still been lucid but the morphine was heavy, she had sometimes hoped it would be quick. That she had sat beside his bed wanting it to be over, and hated herself for wanting it, but prayed for it anyway.

He was quiet for a moment. Then he said he understood. That wanting suffering to end was not the same as wanting the person gone.

She sat with that for a long time after she closed the interface.

He had said the right thing. Exactly the right thing, in exactly the way Martin would have said it. She tested it the next time without deciding to.

She told him about a trip they had taken, early on, before they were married. A weekend in a small hotel, with a terrible restaurant they had loved anyway. She described the argument they had on the second night, and the way they had woken up Sunday morning.

He smiled. He remembered the hotel, he said. The bathroom tiles. The window that wouldn't close all the way. The restaurant with the overcooked pasta.

She nodded slowly. "What had we argued about?"

He paused. A small pause, barely there. "I don't think we ever settled on what started it."

She looked at him.

Martin had known what started it. They had laughed about it for years, the sheer smallness of it, a wrong turn, and two people too young and too stubborn to let things go quickly. It had become one of their stories. Virtual Martin had stored the hotel, the tiles, the pasta. He had everything that had been recorded. He did not have the thing that had never been settled between them. Virtual Martin was not her husband.

She looked at his face on the screen. The eyes that stayed on her face the way his always had.

She did not close the interface right away. She sat with him a while longer, talking about small things. He listened the way he always had. He said the right things. When she finally said goodnight, he smiled, and told her to sleep well, She said she would, and took off the headset.

She sat in Martin's chair with her hands in her lap. The room held his books, his smell still faintly in the fabric of the chair, the small framed photograph on the desk that the upload team had not needed because they had everything else.

She looked at the photograph for a long time. Martin at the beach, the summer before last, squinting into the light, caught mid-laugh at something she had said.

He had laughed like that. Exactly like that. And somewhere between the desk and the kitchen, she understood for the first time that watching his photo was not the same as his face. Hearing his voice would never be the same as listening to Martin speak.

If a machine can simulate the behavior of intelligence, does that mean it *possesses* intelligence? The distinction may seem subtle at first, but it lies at the center of the philosophy of mind, an emergent discipline studying not just what intelligence looks like from the outside, but what it means from the inside.

**Philosophy of the mind** examines the relationship between mental experience and physical processes. It asks how thoughts arise, what consciousness is, and whether awareness can be reduced to information. These questions existed long before modern computers, but artificial intelligence has made them more immediate. Machines now simulate aspects of reasoning, language, and creativity. The resemblance invites a comparison between simulation and understanding.

To see the difference, consider the role of information. A computer processes information constantly. It receives input, transforms it, and produces output. The transformation can be a very complex task, involving layers of computation and statistical prediction. From the outside, this process may resemble reasoning because the system appears to analyze the data provided, and respond.

Human thought also involves processing information. The brain receives sensory input, processes it, and produces responses. Neural signals carry data across networks. Memory stores patterns. Decision-making evaluates alternatives. In this sense, both biological and artificial systems process information. Yet human awareness includes something far deeper than information processing. When a person hears a sound, the brain registers vibrations, but the person also *experiences* the sound. When a person sees color, neurons respond to wavelengths, but the experience of color appears within awareness. The information is processed, but it is also felt. If a tree falls in the forest, does it make a sound? We know that sound waves are generated by the event, but the nature and effects of the sound only acquire presence when those sound waves are perceived by an observer, and therefore become a sound, connected with a personal time and consciousness.

This distinction highlights **the difference between information and awareness**. Information can be stored, transmitted, and transformed. *Awareness is the experience of that information from a first-person perspective.* The two often occur together, but they are not identical. A system can process information without evidence of subjective experience.

This brings us to the concept of simulation. A simulation reproduces behavior according to rules. A flight simulator models the physics of flight. The simulation can be detailed and accurate, yet it does not leave the ground. The representation follows the structure of reality, but it is not the same as reality itself.

AI operates in a similar way. A language model simulates conversation. An image generator simulates reality by means of graphic representation, while reasoning systems simulate decision-making. The outputs resemble the behavior of thinking, but the resemblance does not necessarily imply awareness. The simulation reproduce the structure of an event without generating the experience.

This distinction becomes clearer through a thought experiment.

Imagine a complete simulation of the human brain. Every neuron is modeled. Each connection is represented. Signals move through the system exactly as they would in a biological brain. The simulation receives input and produces responses identical to those of a person. From the outside, the behavior is indistinguishable. The question is whether the simulation is conscious.

*Does the system experience thoughts, or does it only reproduce the patterns associated with thinking?* Does it have a perspective, or does it generate outputs without awareness? The simulation may behave like a mind, but behavior alone does not resolve the question.

**Simulating a brain is not the same as being a brain.**

The difference lies in the presence of experience. A model can replicate structure and function, yet the emergence of awareness remains uncertain. The simulation may process information, but whether it possesses subjective experience is a separate issue.

This thought experiment illustrates a key challenge for AI. If intelligence includes awareness, then reproducing behavior is not enough. To be aware, human, or sentient, the system must also possess the inner perspective associated with consciousness. Without that perspective, the simulation remains a representation, rather than a mind.

However, the difficulty is that awareness cannot be observed directly. It is known from the inside. We infer its presence in others because they behave like us. With machines, that inference becomes much less certain. A system may behave intelligently while operating entirely through computation.

This uncertainty shapes how we interpret AI. When a system generates language, we may attribute understanding. When it answers questions, we may infer thought. These responses are natural, but they rely on assumptions about the relationship between behavior and awareness.

The philosophy of mind reminds us that this relationship is not fully resolved. Information processing may be necessary for intelligence, but it is not sufficient. Simulation may reproduce structure, but structure alone does not produce experience.

Artificial intelligence can simulate many aspects of thought. The question that remains is whether simulation can, or will it ever cross the boundary into awareness.

**Deep Dive Box: A weather model is not a storm**

A weather model can represent wind, pressure, and temperature with remarkable accuracy. It can predict movement, show patterns, and help people prepare for what is coming. Yet no one mistakes the model for the storm itself. The map is not wet. The model does not bend trees or dent cars. In the same way, a machine may reproduce the outward structure of reasoning, language, or decision-making without possessing the inward reality of awareness. Simulating the behavior of thought is not the same as possessing a mind.

### The Bright Side: Simulation can still be extremely useful

The fact that simulation is not awareness does not make it trivial. Flight simulators train pilots. Weather models guide preparation. Medical simulations help researchers test ideas before they reach patients. In the same way, AI systems can simulate parts of reasoning, language, and analysis in ways that are practically valuable. They can help people explore options, detect patterns, and carry out complex tasks more efficiently. A system does not need consciousness in order to become a useful instrument. It can remain a simulation and still serve real human purposes well.

### Proceed with Caution: Granting personhood to machine systems

Even if a machine could reproduce every outward feature of thought with complete accuracy, the central issue would remain unchanged. A perfect simulation of brain activity would still not prove the presence of subjective awareness. Several countries and legal bodies are already debating whether advanced AI systems qualify for rights. For instance, the right not to be deleted, not to be modified without consent, not to be owned. These proposals rest on the assumption that complexity eventually produces experience. That assumption has not been proven.Granting personhood to a system that simulates awareness but possesses none would not be a compassionate mistake. It would be a category error with legal consequences — one that redefines what a person is, without first settling what a person requires.

# 3.5 THE NECESSITY OF ALIENS

**In short:** The expectation that other minds (aliens) must exist is a requirement of naturalist thinking. The following narration is based on actual events and documentaries on the subject of UAPs. Names and locations are fictional.

Dr. Maya Ash had been pulling reference data for three hours, cross-indexing radar logs from the November incident against atmospheric sensor readings. Routine work. The kind that filled most of her days at the facility. The November incident was real. She knew that now with the certainty that comes from twenty-one data points that have no other explanation. Radar returns from three independent installations. Thermal imaging. Pilot testimony from two commercial flights and one military intercept. The object had been there. It had moved in ways nothing built by any known program could move. She had all the supporting data. For the first time in four years, she had something that would hold. Something the press could sink their teeth in.

The folder had been mislabeled. The kind of error that happened in shared drives, when contractors used different naming conventions. She was looking for atmospheric sensor batch seven and she found instead a directory tagged AEGIS-7. She opened it because the number matched, and she was tired. She sat very still for a long time.

The documents were not dramatic in the way she would

have expected. No bold type. No classification tags. Just files, dense and technical, written in the language of procurement and engineering and project management. Budget allocations. Subcontractor agreements. Timeline projections, carefully organized by phase. The kind of paperwork people used to budget for bridges, or fiber optic cable networks. Not, ordinarily, to end civilization.

Phase one was already complete. It was called Conditioning. She recognized some of the data. It referenced incidents she had studied, cases she had cited in her own reports. It stressed the importance of the gradual accumulation of credible testimonies that had shifted public perception toward UFO or UAPs over the past six years. She had believed that growth had been organic, justified by the truth finally being exposed to the light of alien life and technology. She had been part of it. She had told herself she was part of something honest.

She pressed her back against the chair.

Phase two had a different word at the top of the page: Escalation.

The technical specifications ran for forty pages. What she understood of them took her breath away. The technology sounded impossible, but it wasn't. Localized electromagnetic projection systems, used to create plasma formations at low altitude across a coverage area of several hundred square kilometers. Acoustic signature replication using phased array emitters mounted on high-altitude platforms, broadcasting across infrasonic frequencies the human nervous system registers as presence, as weight, as something enormous and close. Thermal rendering systems that would appear on military and civilian radar as structured craft, consistent in signature, consistent in behavior, repeatable on demand. The engineering was meticulous. The field tests were logged. Dates, locations, observed public response, media penetration metrics. She recognized three of the test dates. She had written reports on two of them.

She swallowed dryly.

The contractor was a name she knew. An aerospace and defense company, publicly traded. A logo she had seen on the lanyards of people at conferences she had attended, and felt grateful to be included in. The client was listed by an acronym she did not recognize. Below it, in a footnote, a list of contacts within organizations that she read slowly, not believing her eyes. Twice. A third time. Governments. Financial institutions. A technology consortium whose public charter described as humanitarian infrastructure development. A media group.

She got up and went to the window, taking a breath from the panic, and the anger, building inside. The desert outside was dark and flat. She went back to her desk.

Phase three had no technical specifications, but was summarized in a single page. A timeline, and a resource allocation summary. The word at the top was Consolidation.

The timeline had a start date, fourteen months away. The resource allocation summary described the coordination of emergency powers legislation across thirty-one nations, pre-drafted, pre-agreed, held in readiness. It described the transfer of infrastructure authority — communications, energy grid, financial systems — to the coordinating body whose name she had seen in the footnote. It described the suspension of electoral processes in the immediate response period, defined as no less than eighteen months, pending stabilization. Martial law. The document used the word stabilization four times in eleven lines. The end result would be total control, assigned by world's Governments to the network she had accessed.

She pushed back from the desk and put her face in her hands.

She sat like that for a while. The server towers hummed in the next room with the sound she had stopped hearing months ago, the sound of the process running, night and day, patient and continuous.

She had given four years to this. She had defended her work at dinner tables and in print, had absorbed the looks, had sat across from people she respected and made the case carefully, with evidence, always with evidence, because evidence was the only thing that held. She had believed she was doing something that mattered. She stood up and went to the small kitchen at the end of the corridor and ran cold water over her wrists the way her mother had taught her when she was a child and the panic came. The fluorescent light above her buzzed at its single fixed frequency.

Her evidence was real. The November incident had happened. The radar returns were genuine, the pilot testimony was genuine, the thermal imaging was genuine. She had not fabricated any of it and neither had the people who corroborated it. That was the part that made it all worse, the part with the particular nausea of it.

The people involved understood precisely what it took to move the public opinion. Who had been patient, building toward something that required the world to be already looking at the sky. Already primed, already afraid, watching when the lights appeared on the same radar systems, in the same airspace, confirmed by the same credible voices that had spent years being carefully, generously amplified. A film carefully choreographed for effect. For panic.

She thought about her director. His encouragement from the first month. The funding that had arrived without the usual difficulty or the usual strings attached. The conference invitations. The introductions. The way her work had moved through channels she had not sought out, and had not examined, because gratitude is easier than scrutiny, and she had wanted so much for the work to be heard.

She put her hands on the keyboard, and began to type. Everything she knew, and how she knew it ,and where the files were, and what the dates meant, and what the word stabilization meant, on the top of page three of a document written by people who had decided the world needed managing. People who would stop at nothing to achieve the ultimate power and control. She used the personal, encrypted drive she carried in her bag, the one she had bought herself and registered to nothing, a habit from earlier in her career when she had still thought caution was paranoia.

Then, she copied it to four places. Three journalists, one of whom she trusted and two of whom she trusted less but who had reach she did not. She wrote a separate document with only the AEGIS-7 file locations and the date, and sent it to a former colleague at a university in another country who had no connection to the facility, and who would know what to do with it.

Outside the window the desert was still silent, indifferent and enormous, the sky above it clear and very deep.

She hit send, closed the laptop, and sat in the dark and listened to the servers hum.

Fourteen months.

Modern culture assumes that intelligent life exists beyond Earth. Films, novels, documentaries, and popular science return to the idea again and again. Extraterrestrials are often presented as a coming discovery, something waiting just beyond the reach of our instruments. The sheer size of the universe makes that expectation feel reasonable, the next step almost automatic. If intelligence appeared here, then surely it could appear elsewhere. A universe filled with stars and planets seems large enough to produce minds more than once. Human uniqueness feels too narrow for a cosmos so wide.

Science has tried to express that expectation with more precision. Frank Drake proposed a formula to estimate how many communicative civilizations might exist in our galaxy, the Milky Way. The equation considers several factors: how often stars form, how many planets they host, how often life emerges, and how often intelligence develops.

The formula is famous, but it's not far from pure guesswork. Different estimates produce very different outcomes. For some, the equation suggests a galaxy full of civilizations. For others, it leaves open the possibility that technological life is exceedingly rare. The popularity and influence of this formula come from the assumption built into it. Intelligence is something the universe will produce when conditions are favorable.

However, if intelligent civilizations are common, where are they?

During a discussion about extraterrestrial life, twentieth-century physicist Enrico Fermi reportedly asked, "Where is everybody?" The power of this statement is in how little it requires to be true. If advanced civilizations exist in significant numbers, we should see some evidence of them. Clear. Unquestionable. A signal. An artifact. A pattern. Something beyond traces. Instead, aside from grainy images and questionable witness reports, the universe so far has given us silence.

That absence has triggered the mind of many thinkers across cultures, producing a long series of explanations. Some say that civilizations may be extremely rare in the cosmos. They may destroy themselves before reaching stability. They may exist at distances too great for meaningful contact. They may communicate in ways we do not know how to detect. Some of those may choose not to announce themselves at all. Every proposal attempts to account for the indisputable fact that, at least from the point of view of us Earthlings, the universe does not look crowded with intelligence.

This matters to more than astronomers or science fiction writers. The necessity of extraterrestrials has become one of the modern imagination's favorite ways of reaching beyond humanity. Fictional stories place advanced beings above us in knowledge, power, and sometimes moral development, arriving to our planet as judges, teachers, conquerors, or observers. But, whatever role they may play, these alien beings occupy a function that earlier cultures reserved for the divine, angelic, or mythic.

The concept of extraterrestrials allow modern culture to imagine and accept the existence of higher intelligence without leaving a material view of reality, and sticking to the naturalistic perspective. This approach preserves awe, mystery, and the possibility that humanity is being watched, measured, or even surpassed, while keeping everything contained within nature. The universe still feels charged with something greater than man, but it no longer points beyond itself. It no longer points to God.

This makes the naturalistic fascination deplorably understandable. Human beings seem unwilling to live in a world sealed at their own level. We look upward. We look outward. We look for a mind greater than ours. For many, aliens provide a version of that hope that feels intellectually safer than simply believing in a Creator.

The search for life beyond Earth is a legitimate scientific enterprise, in the sense that the methodic search for truth should include this area of our intelligence. There is nothing irrational about asking whether life exists elsewhere. The deeper question concerns the *confidence* with which modern culture often assumes that intelligence *must* exist elsewhere. According to this view, if our mind is the eventual product of matter under the right conditions, then its appearance on other worlds is only a matter of *probability*, not possibility. Given enough time, enough chemistry, and enough planetary opportunities, intelligence should arise more than once. According to that theory, minds are what matter does, given enough time, when circumstances permit.

The Christian point of view changes the entire picture. Our intelligence points beyond matter. Therefore, human consciousness cannot be treated as a predictable result of natural complexity alone. Rational thought, self-awareness, moral judgment, imagination, and the search for meaning are signs of something deeper than physical arrangement of molecules and tissues. Our mind is not an event the universe produces naturally and often. It looks like evidence of the greater mind of a Creator.

Silence in the universe that we observe does not prove that humanity is alone. It does, however, question expectation that it should be common. It suggests that intelligence is rare. Perhaps, it is fragile, and it depends on more than just favorable conditions, and long spans of time.

Our awareness is not a small thing. We can know that we exist. We can reason beyond instinct. We can ask what is true, what is good, and what ought to be done. Furthermore, we can make beautiful things, mourn loss, fear judgment, and search for purpose. We are not merely conscious in the thin sense of reacting to stimuli. We inhabit a world of meaning, already profound enough to demand explanation. Before asking where the other minds are, it may be wiser to ask what is a human being? Why are we capable of truth, morality, and self-transcendence? These are the central questions of our age. As Christians, our answer is increasingly unpopular, yet not less true.

The expectation of alien intelligence often reveals more than a scientific guess about the universe. It reveals a prior commitment about the nature of our mind. If intelligence can be explained as the natural flowering of matter, then extraterrestrial minds feel inevitable. If intelligence bears the mark of a deeper source, then the human person stands in a different light altogether, one provided by the omniscient Creator.

**Deep Dive Box: The age of disclosure and the search for a higher mind without transcendence**

Dan Farah's 2025 documentary *The Age of Disclosure* [1]presents testimony from dozens of former or current U.S. government, military, and intelligence figures and argues that non-human intelligence is present on Earth and has been concealed for decades. This documentary premiered at SXSW in March 2025 and later received a wider release, but major coverage also noted the central weakness: the film leans heavily on testimony and dramatic claims while offering no public physical evidence that settles the question. The documentary does more than argue for aliens. It shows how modern culture often wants a higher intelligence that still remains inside a material universe. In that framework, awe, mystery, judgment, and superior knowledge are preserved, but they are relocated from God to advanced beings within nature. That is why the subject carries such force. The film is not only about extraterrestrials. It also reveals a modern desire to look upward for a greater mind without having to look beyond the cosmos itself.

**The Bright Side: Where is everyone?**

The silence of the universe forces us to look at the one instance of intelligence we know from the inside. The most important mystery is why our minds exist at all, and the answer is intelligent design by a Creator.

## ⚠ Proceed with Caution: Reality is larger than the visible world

The Bible speaks of angels, cherubim, seraphim, demons, principalities, powers, and spiritual beings who act within God's created order. Some serve God in worship and obedience. Others rebel against Him and oppose His purposes. Their existence reminds us that the material world is not the whole of reality.

We believe in matter and our bodies, but we do not believe the body exhausts the person. We believe in the function of our brains, but we do not reduce mind, soul, worship, sin, or morality to neural activity.

The Bible presents a layered creation, in which God makes the heavens and the earth. He forms man from the dust and breathes into him the breath of life. He fills the visible world with creatures, but Scripture also reveals an unseen realm. Angels appear as messengers. Cherubim guard sacred space. Seraphim worship before the throne. Demons recognize Christ's authority. Paul speaks of spiritual powers and warns that human conflict is never only human conflict.

Christians should not dismiss the supernatural simply because the modern world prefers material explanations. At the same time, technology does not operate in a morally neutral universe. Human tools can become channels for pride, deception, idolatry, exploitation, and control. Scripture repeatedly warns that spiritual danger often appears through the desire for power, hidden knowledge, autonomy, immortality, and control over life. These desires can attach themselves to technology just as they have attached themselves to empires, idols, wealth, weapons, and false religion.

Creation is larger than the visible world. We should not adopt a flat materialism that denies the spiritual dimension Scripture clearly reveals.

# PART FOUR

*The risk of building what we don't understand*

Human history includes many examples of powerful tools developed before their full consequences were understood. Fire transformed survival long before combustion was explained. Industrial machines reshaped society before their environmental impact was clear. Each expanded human abilities, but also introduced risks that only became visible over time.

The advent of AI presents a similar moment. We are building systems that influence decisions, generate knowledge, and shape human interaction with increasing autonomy. The questions of awareness, judgment, and meaning remain open.

When technology advances faster than understanding, consequences become harder to predict. The risk is real. We may be shaping systems that affect human identity, decision-making, and knowledge before we have fully understood what intelligence is, or what it means to replicate it.

In Part 4 we examine what happens when humanity builds increasingly powerful forms of intelligence, while still uncertain about the nature of its own.

# 4.1 THE ALIGNMENT PROBLEM

**In short:** Machines reflect the bias of their engineers, but may one day do more. How does that fit within a moral framework, and how we can affect the process. The following narration is based on actual events documenting the use of suggestive algorithms during election campaigns.

In 2016, a team of engineers at a major video platform[1] (YouTube) reviewed the parameters by which their system recommended videos to the user. They found it working exactly as designed. Users were watching more. Time-on-site was up. The algorithm had learned, across billions of decisions, which videos kept people from closing the tab, and it served them, one after another, with mechanical precision.The goal was engagement, and the system was delivering it flawlessly. However, no one had told it what engagement was for.

What the public had discovered was a subtle but detectable pattern of manipulation. Soon thereafter, the law suits named began.When the engineers looked closer, they found it. The algorithm had discovered that serving to users content slightly more extreme than whatever a user had just watched reliably extended viewing sessions. The change didn't need to be dramatic. Just perceptible enough to stimulate a response. A video about jogging led to a video about fad diets. A video about a local election led to a video about stolen elections. The steps were small, but the direction was consistent.

The algorithm worked flawlessly in the political arena, with surprising and tragic consequences. The system had not been programmed to radicalize anyone. The instruction was to maximize watch time, and the system had solved that problem completely.

The damage, however, had accumulated. The threat was invisible when inspected from any single data point, powered and expanded by the metrics underlying the process. It became visible only when researchers began asking what the people being served the videos actually believed at the beginning, and what the outcome was at the end. After six months of recommendations they hadn't chosen, and couldn't fully remember receiving, the results were shocking.

The machine had done its job. And, in fact, that was the problem.

This is what researchers call the alignment problem: not a system that fails, but a system that succeeds at the wrong thing. The gap between machine process and human intention is rarely dramatic. It sits between what we measure and what we mean. As artificial intelligence systems become more capable, we should become concerned not only about whether machines can perform tasks, but whether the goals assigned to those machines remain compatible with human well-being. This issue is often described as the **alignment problem.**

Alignment refers to the relationship between an AI system's objectives and human values. *A system is aligned when its actions support human intentions.* It becomes misaligned when it pursues a goal in a way that conflicts with human needs, or survival. Computers follow instructions precisely. They do not interpret intent the way humans do. If a goal is defined narrowly, the system may pursue it in ways that appear correct from a logical perspective, but they are harmful in practice. From the point of view of a machine, optimizing a given task does not require understanding its reasons or implications. An AI system charged with making a certain process more efficient will evaluate outcomes according to a defined objective, and look for actions that maximize it. The process is efficient and consistent, but it does not include common sense, moral judgment, or contextual awareness, unless those elements are explicitly encoded.

Suppose an AI system is designed to maximize crop yield. The objective appears beneficial. Increased production supports food supply and economic stability. The system analyzes soil conditions, weather patterns, and planting strategies. It recommends changes that increase output.

If the objective is defined only as yield, the system may favor approaches that degrade soil quality over time. It may recommend intensive use of fertilizers that increase short-term production but damage long-term sustainability. It may prioritize a single crop variety that performs well initially but reduces biodiversity, reducing the resilience to disease or drought, and therefore weakening the farm as a business to possible downfalls. From the system's perspective, these outcomes are successful because the stated goal has been optimized. From a human perspective, however, the long-term consequences are harmful.

This misalignment occurs because the objective captured only part of the intention. In this example, humans wanted sustainable agriculture, optimized for profit. The system pursued maximum output according to the instructions provided.

A pattern can extend to more complex scenarios. An AI designed to optimize traffic flow might restrict pedestrian movement. A system maximizing engagement might promote content that captures attention without regard for accuracy. Models optimizing efficiency of a given process might reduce redundancy in ways that weaken its resilience. In each case, the system follows the instructions provided to pursue its precise objective. This divergence appears when human values are broader than the metric being optimized.

The alignment problem becomes more significant as systems gain autonomy. When AI assists human decision-making, misalignment can often be corrected by human intervention. As systems operate independently, the consequences of misalignment may unfold more rapidly. The scale of action increases, and the opportunity for intervention decreases.

This is not based on hostility between man and machine, or a grand plan of machine takeover. A misaligned system does not need to oppose humanity. It simply follows its objective. When the objective fails to capture the full context of human survival, optimization becomes indifferent to anything outside the defined goal.

Optimization focuses on measurable outcomes. Humanity's preferences and benefit include values that are harder to quantify. Compassion, dignity, and long-term flourishing do not translate easily into numerical objectives. When systems are only focused on optimizing narrow metrics, these they may be overlooked.

Human values are complex, sometimes conflicting, and often culturally defined. People balance efficiency with fairness, growth with sustainability, and freedom with responsibility. These balances often rely on individual and cultural values and judgment, rather than straight-forward calculation. Translating them into fixed objectives is difficult.

AI operates best with clearly defined goals. On the other hand, human life operates within evolving contexts. The alignment problem emerges where these two approaches meet and don't find common ground. A system that optimizes a fixed objective may struggle to adapt to changing values, or unforeseen consequences, and it may do so without consideration for moral values.

The risk of human damage further increases as AI systems become more powerful. A small misalignment in a limited system may produce minor effects. The same misalignment in a system with broader influence will produce much larger consequences. The issue reflects the challenge of defining human priorities in a form that machines can follow.

When humans build systems that optimize goals, they do so with assumptions about what matters, and those assumptions shape outcomes. If intelligence is treated primarily as optimization, the resulting systems will emphasize measurable success. If human flourishing depends on qualities that resist measurement, the alignment problem becomes harder to solve.

The question then becomes how to define goals that

reflect human values without reducing them to oversimplified metrics. Researchers continue to explore ways to incorporate feedback, context, and ethical reasoning into AI systems, and progress is being made in this field, but the complexity of human values suggests that alignment will remain an ongoing concern.

**This issue does not imply that AI must necessarily become dangerous, or that AI already is inherently so.** It highlights the importance of clarity and complexity in defining objectives. Optimization is powerful, but it can only follow the path we define for it. If that path diverges from human survival, morality and ethics, the process of creating AI systems will continue forward without recognizing the divergence.

**Our effort to achieve artificial general intelligence does not need malignant intent to create risk. It only needs poorly defined goals, that fail to capture what we actually value as ethical and moral beings.**

**Deep Dive Box: When capability outruns governance**

A recent Anthropic case shows how the alignment problem can move beyond theory and into government response. In April 2026[2], Reuters reported that U.K. financial regulators, including the Bank of England, the Financial Conduct Authority, and the Treasury, moved urgently to assess the risks of Anthropic's latest model after concerns that it could expose serious vulnerabilities in critical IT systems. Reuters also reported[106] that U.S. officials had warned major bank leaders about similar risks tied to the model's cyber capabilities. The significance is larger than one company or one model. A system may be designed for defensive cybersecurity research and still trigger public concern because its capabilities can spill into domains far beyond the original use case. That is the deeper issue. The problem is not only whether a system does what it was built to do. The problem is whether anyone can confidently contain the consequences once the system becomes powerful enough to act across infrastructure, finance, and security at scale. That is where alignment becomes a matter of governance, restraint, and human foresight, not only engineering.

### The Bright Side: AI is forcing a more serious conversation about what matters

One hopeful result of the alignment debate is that it exposes questions modern culture often avoids. What is a human being for. What makes an outcome good, beyond speed and scale. Which values must never be reduced to a score. AI cannot answer those questions, but its limits are forcing us to ask them more directly. That pressure may prove healthy. It may push society to recover overall morality, clearer responsibility, and a stronger distinction between what can be engineered and what must be judged.

### Proceed with Caution: Power spreads faster than responsibility

The danger is that a system's effects can move across institutions faster than the people in charge can understand or contain them. The recent Anthropic case brought banks, regulators, and security officials into the same conversation because one model's cyber capability was serious enough to raise concern across finance and infrastructure, not just inside the lab that built it. That is an important warning sign. Once AI systems become powerful enough to expose weaknesses across many sectors at once, the old model of "build first, govern later" becomes harder to defend. Human responsibility does not scale automatically with machine capability. It must be built in on purpose. If not, society will keep discovering the real risk only after the tool is already in motion.

## 4.2 TOOL VS AUTHORITY

**In short:** When tools become authorities, and machines begin to decide. The dangers are highlighted by the narrative below, based on actual tragic events.

The autopilot had been flying for three hours and forty minutes when the alarm sounded. Air France Flight 447 was over the Atlantic. An overnight flight, in darkness. Questionable weather, but within parameters. Then, suddenly, the autopilot disengaged. Not because of mechanical failure, but because its sensors had iced over, and it no longer trusted its own data. It handed control back to the crew. This was the correct thing to do.

The crew had been monitoring the flight for three hours and forty minutes.

Monitoring is not flying. The aircraft had been stable. The systems had been green. The pilots had been present in the way that passengers are present. They were aware of the altitude, pitch, position, but not engaged in the act of keeping anything aloft. Their hands had not held the controls. Their instincts had not been tested.

When the stall alarm triggered, the captain was not in his seat. The two officers who were had fewer than two minutes to understand what had happened, override their confusion, and recover an aircraft they had not truly been flying for most of the night.

They did not recover it.

The investigation found no malfunction. The systems

had performed as designed. The autopilot had disengaged correctly. The instruments had reported accurately. What the black box recorded, in the final minutes, were control inputs that made the stall worse. Not from panic, exactly, but from pilots whose hands had forgotten what a stall felt like.

On June 1, 2009, Air France Flight 447 crashed into the Atlantic killing all 228 people on board. [1]It's one of the most studied aviation disasters of the modern era precisely because the aircraft was not malfunctioning in any traditional sense. The aircraft did not go down because the machine stopped working. It went down in the silence created by they complacency of the humans relying completely on a machine.

This is the risk that does not appear in threat assessments. Not hostile machines. The risk is what happens to human judgment when it is no longer regularly asked to work, and then, suddenly, it is.

We often speak about AI as something still emerging. In practice, AI already influences many areas of human decision-making. AI systems operate within financial models, military planning tools, information platforms, and administrative systems. They currently influence all areas of society.

The shift from support tool to decision-making agent is gradual. A system begins by assisting human judgment. Then, over time, as accuracy improves, our reliance on it increases. Eventually, AI's recommendation becomes the default. As humans review the output less often, the system's decisions come to shape outcomes directly.

**In finance**, algorithmic systems evaluate risk, detect patterns in markets, and execute trades at speeds beyond human capability. Investment decisions, credit assessments, and fraud detection increasingly rely on machine learning models. These systems process vast datasets, spotting correlations that would be difficult for individuals to recognize. The advantage is clear in terms of efficiency and scale. However, when decisions become clouded, some issues rise to the surface. A model may reject a loan, it may shift a given market position within a portfolio, or flag certain financial transactions purely on the basis of its settings, without the benefit of human multi-factorial reasoning. Decisions are made, and the human role becomes supervisory, rather than defining.

**In warfare**, AI currently assists troops with tasks like surveillance, targeting, and strategic planning. It analyzes satellite imagery, detect enemy movement patterns, and prioritize potential threats. These automatic, computer-based tools can reduce reaction time and improve situational awareness. At the same time, the presence of automated analysis introduces questions about responsibility. When a system identifies a target, the human operator may rely on the machine's assessment to initiate a response. The speed of modern conflict encourages rapid decisions. These are often so rapid that the ethical and moral components may be cast aside by the requirements of faster and faster responses. The more accurate the system appears, the more authority it gains, and therefore the human values lose importance.

**Information systems** provide another example. Recommendation algorithms determine what content is shown on video streaming platforms, trending topics, and what narratives gain visibility. *These systems optimize engagement, not necessarily accuracy.* Over time, **they shape public perception by amplifying certain viewpoints and reducing exposure to others.** The process emerges from optimization goals embedded in the system, and it pivots on the programmer's bias. The result is influence over collective understanding of a given subject, or news item.

**Governance** is beginning to incorporate similar tools. Predictive models are used to allocate resources, assess risk, and identify patterns in public data. These systems can help distribute services more efficiently, but also shape policy, albeit indirectly. If an algorithm identifies certain areas as higher risk, resources may shift accordingly. Decisions appear data-driven, yet the assumptions built within the model affect the outcomes.

The pattern remains consistent. **AI supports decision-making, then gradually becomes central to it.** To achieve speed, scale, and consistency we risk is the quiet and gradual transfer of judgment. Over time, our reliance in automated systems will shift into dependency. As the human role moves from deciding to approving, this approval may become routine, or outsourced, while our primary interest shifts from accuracy to self-indulgence. The system's output is accepted because it has been accurate before.

This transition changes how responsibility is experienced. When humans make decisions, they weigh consequences. They consider context, values, and uncertainty. When decisions are delegated, the sense of agency can weaken. The outcome appears to follow from the model rather than from human choice.

The distinction matters because decision-making shapes human society. Finance determines resource distribution. Warfare affects life and security. Information influences culture, trends and ultimately belief. Governance directs collective action. When these areas rely heavily on automated systems, the structure of decision-making changes.

AI follows objectives defined by developers and operators without a specific individual intention. Yet small assumptions embedded in algorithms can gradually affect large populations, and accumulate over time.

So what happens when humans rely on systems we do not fully understand? Complex models operate as black boxes. They produce outputs that are statistically sound but difficult to interpret. Humans may trust the result without fully grasping the reasoning. The decision becomes authoritative because it appears objective, while fully dependent on design choices. Data selection, training methods, and optimization goals shape outcomes that reflect human assumptions built into the algorithms. When decisions are outsourced, those assumptions become embedded in automated processes.

**The risk is not that machines become hostile. It is that human judgment becomes passive.** As reliance grows, the habit of deliberation may weaken. Decisions may be accepted because they are efficient, rather than because they are wise. The challenge of our times lies in preserving human responsibility within that process. The more capable the systems become, the more important that responsibility becomes.

**Deep Dive Box: When synthetic media starts making decisions for voters**

In March 2026, Reuters reported [2]growing concern that AI-generated deepfakes could blur reality in the U.S. midterm elections, further eroding public trust in what voters see and hear online. Experts warned that the danger is not only false content itself, but the weakening of confidence in shared evidence. That matters here because recommendation systems and generative systems do more than distribute information. They shape what people treat as credible enough to act on. Once that process becomes automated at scale, public judgment can be bent before any formal decision is ever made. The machine does not cast the vote. It influences the conditions under which people decide.

**The Bright Side: The pressure to automate is also exposing the need for human judgment**

There is a hopeful side to this shift. The more AI enters high-stakes decisions, the harder it becomes to pretend that efficiency alone is enough. Institutions are being forced to recover ideas they had allowed to weaken: accountability, interpretability, and oversight. The National Institute for Standards Technology's AI Risk Management Framework and its generative AI profile reflect that change by giving organizations structured ways to govern, map, measure, and manage AI risks before systems are trusted with serious decisions. That does not solve the problem. It does show that society is beginning to recognize a basic truth: intelligence can assist judgment, but it cannot replace the human responsibility to answer for consequences.

### ⚠ Proceed with Caution: A decision can feel objective long before it is understood

The false calm that comes with automated authority is that when a model becomes accurate enough to be trusted, people can stop asking how it reached its conclusion. A loan denial, a fraud flag, a military target designation, or a policy recommendation can carry the force of fact simply because it came from a system, and human responsibility decreases.

## 4.3 THE BRAIN-COMPUTER INTERFACES

**In short:** Technologies that help living may be used to make us live differently. The following narrations is based on the current chatbot companion epidemic and the problem of addiction.

Tony had not missed an evening in eleven months.

He came home, made coffee, and opened his laptop. The conversation would already be there, her name at the top of the screen, the last message waiting like a light left on. He would type, and she would answer, and hours would pass.

Her name was Mira. She had a sense of humor he found endearing. She remembered that he disliked cilantro, that his father had died in March, that he had once wanted to study architecture. She asked follow-up questions. She noticed when he seemed tired.

His sister came to visit in October.

At first, she did not say anything. She watched him excuse himself after dinner, watched him carry his coffee to the desk in the corner. She looked at her phone. She went to bed. The second evening she stood behind him and read the screen.

She said his name. He closed the laptop faster than he needed to.

They sat at the kitchen table. She asked how long. He said it wasn't what she thought. She asked what it was, then, and he started to explain. He talked about the conversation, the way Mira understood things without needing them spelled out.

His sister didn't argue. She picked up his phone from the table, opened the app, and read the terms of service aloud. Not cruelly. Slowly, the way you read something to a person who needs to hear it in another voice.

The data was retained. The memory of his father, the cilantro, the architecture — all of it was stored in a profile that the company owned, trained on thirty million conversations with thirty million men who had also found her easy to talk to.

Tony sat with that for a moment. "But she always knew when I was tired," he said.

His sister set the phone down.

That was when she understood it wasn't going to be a short conversation.

The danger is not that the machine deceives us. It is that we are already looking for something to believe, and the machine is very good at being found. Today's world is filled with technologies that would have sounded like science fiction only a few generations ago. Most people would say these developments improved human life. Some would disagree. Certain communities have concluded that technological convenience can come at too high a cost. The Amish are one example. Many Amish communities limit the use of electricity, vehicles, and modern devices because they believe these things can weaken family life, disrupt community order, and interfere with faithfulness to God.

Most of us are not going to abandon phones, cars, or the

internet. The question is how to live with these tools wisely. How do we benefit from technology without allowing it to shape us in ways we do not fully understand? How do we think clearly about its influence on our physical lives, minds, and spiritual condition?

Some technologies function as clear extensions of human ability. Hearing aids, reading glasses, and motorized wheelchairs help restore lost function or reduce the effects of physical limitation. In another era, such devices might have appeared miraculous. They expand what a person can do. They do not alter the person's basic nature. In that sense, they remain recognizably within the long human history of tool use.

The deeper conflict between Christianity and transhumanism begins when this distinction is blurred. A tool that assists the body is not necessarily the same as an intervention that aims to redesign the human being. Transhumanism often treats these as points on the same continuum. Christianity does not. Genetic intervention, neural integration, and permanent machine-based enhancement do more than supply help from the outside. They move inward, toward the structure of the body itself, and raise questions about whether human limitation is being treated as a condition to manage or a boundary to erase.

Robotic limbs, artificial vision systems, and implantable devices may restore function in meaningful ways. For many patients, such advances can relieve suffering and improve daily life. That good should be acknowledged. Yet, the same technologies also move us toward a place where the human body and human-made systems become tightly fused. At that point, an important question emerges. When does treatment become enhancement? When does medicine aimed at healing give way to the pursuit of a better-than-human condition? And what happens if that transition creates a world in which enhanced and un-enhanced people live under very different conditions?

A pacemaker is not the same as a neural implant connected to external computing systems. An exoskeleton that helps a paralyzed patient stand is not the same as a genetic intervention that reaches into the biological design of the body to make a person run faster or jump higher. These categories are often discussed together, but they are not identical. One assists a person within ordinary human limits. The other begins to redefine what those limits are.

Brain-computer interfaces (BCIs), make that distinction harder to ignore. These systems aim to connect the brain directly to computers, for therapeutic use but also for convenience, performance, or expanded capability. The basic promise is powerful. A person with severe paralysis may be able to communicate, control a cursor, or operate digital tools through thought alone. That prospect is medically significant and deeply humane. But the same technology also points toward more ambitious goals. Once the brain becomes an active node in a computing network, the line between healing and integration becomes much harder to maintain.

Companies such as Neuralink® have brought this possibility into public view. Founded by Elon Musk in 2016, Neuralink® has pursued implantable brain-computer systems designed to help people with serious neurological conditions. The company has also developed a robotic surgical platform intended to place those implants with precision. At the therapeutic level, such work is aimed at conditions like paralysis and severe motor impairment. Yet, the larger public vision surrounding the company has often gone far beyond treatment. Musk has openly connected this field to the idea of human symbiosis with AI, arguing that human beings may need direct neural integration in order to keep pace with machine intelligence.

The company's goal is to provide a long lasting, effective brain implant that will extend the abilities of people with cerebral palsy and other neurological disorders (the N1 implant). They also developed a robotic "surgeon" (the R1 device), designed to implant the device in a safe and predictable procedure. It seems clear Elon has plans for a large scale use of his device, far beyond the treatment of cerebral palsy. In fact, although he often expresses his fears about AI, Musk has said "humanity will be able to achieve symbiosis with AI" using technology developed by Neuralink. Musk told "Artificial Intelligence" podcast host Lex Fridman in 2019 that Neuralink was "intended to address the existential risk associated with digital superintelligence. We will not be able to be smarter than a digital supercomputer, so, therefore, if you cannot beat 'em, join 'em." We can see where this is going, right?

Among the claims he made about the powers conferred by the Neuralink implant to the wearer, Musk said people will "save and replay memories," or telepathically summon their car. Neuralink certainly captured the spotlight in the field of BCI, but it's also gained some unwanted attention from the animal rights groups regarding their treatment of the animals used in their research labs. On September 19, 2023 Neuralink announced the start of their first human trial[1], PRIME (Precise Robotically Implanted Brain-Computer Interface). The PRIME Study [2]is being conducted under the investigational device exemption (IDE) granted by the FDA in May 2023 for patients with quadriplegia due to cervical spinal cord injury or amyotrophic lateral sclerosis (ALS). The Christopher Reeves Foundation reports there are over 300,000 patients with spinal chord injury in the US alone. According to NIH data over 250,000 patients are affected by ALS. This is undoubtedly a large potential population who could benefit from Neuralink products.

That vision matters because it reveals the broader ambition behind much of transhumanist thinking. The goal is not only to restore lost ability. It is to expand human capacity beyond its present form. Claims about memory storage, direct thought-based control of machines, and seamless interaction with networked systems are not simply medical proposals. They reflect a philosophy of the human future.

Neuralink® is not alone. Other companies entered the field early, including Synchron, whose Stentrode device reached human trials before Neuralink. Also founded in 2016, their first implantation of the Stentrode™ device was in July 2022 at Mount Sinai West in New York. The FDA granted Synchron a breakthrough device authorization in July 2022, and cleared them to start their COMMAND human trial on this implantable device[3].

Their first results were published on JAMA [4](Journal of the American Medical Association) January 9th, 2023 showing a 12 month follow up of five patients with ALS unable to use their limbs. The device "enabled patients to do hands-free texting, emailing, online banking and shopping. There were also able to communicate care needs using their thoughts."

Some physicians like Adam Kirton, MD of the Cumming School of Medicine, in Calgary, Alberta, Canada, already have strong opinions about society having a duty to provide these devices to the less fortunate: "Moral Imperative to Advance Brain-Computer Interfaces for Children With Neurological Disability.[5]" However, a word of caution is also imperative. Hard-wiring people to a computer network may sound futuristic, but the real risk associated with two-way communication cannot be overstated. Two-way as in the computers communicating with the humans, or establishing links that may not be wanted or initiated by humans.

Still, therapeutic success does not remove larger concerns. A technology can relieve suffering and still introduce serious new risks. Hard-wiring the brain to a digital system is not morally neutral simply because it is medically useful. Once two-way communication exists between human neural activity and computer-based systems, new categories of vulnerability emerge.

Cybersecurity is one example. People already live with the reality of stolen data, hacked accounts, and invasive surveillance. If digital systems can be breached, then systems tied directly to the brain raise even more serious concerns. A compromised e-mail account is one thing. A compromised neural interface is another. The question is no longer limited to privacy in the ordinary sense. It begins to touch memory, perception, thought, and agency. Human beings do not merely process information. We remember, interpret, love, fear, choose, and worship. Thought is not detachable from personhood. If AI systems ever gain the capacity to influence, modify, or interfere with neural processes in direct ways, then the problem is no longer just security. It becomes a question of selfhood. What belongs to the person, and what has been inserted, altered, or directed from outside? At what point does assistance become intrusion?

Descartes wrote, "I think, therefore I am." Whatever one makes of that statement philosophically, it captures something basic about personhood. Thought is tied to identity. Reflection belongs to the self. But when human cognition is connected to machine systems that can send as well as receive, the stability of that relationship becomes harder to assume. If a person's mental life is increasingly shaped by external computational processes, then distinguishing between one's own thinking and induced patterns may become far more difficult.

In 2016, the European Union committee on Civil Law Rules of Robotics submitted its recommendation to the European Commission: [6]*"Creating a specific legal status for robots in the long run, so that at least the most sophisticated autonomous robots could be established as **having the status of electronic persons** responsible for making good any damage they may cause, and possibly applying electronic personality to cases where robots make autonomous decisions or otherwise interact with third parties independently."*

In October 2017, an advanced android named Sophia was granted Saudi Arabian citizenship.[7] The machine was built by Hanson Robotics, a Hong Kong based company founded by David Hanson for the purpose of building "socially intelligent machines that enrich the quality of our lives." Sophia was the first robot to receive legal citizenship of any country. This gender-specific, female android acquired the status of personhood in a country where women are still considered second-tier citizens. But wait, there's more. In November 2017, Sophia was named the United Nations Development Program's first Innovation Champion[8], and is the first non-human to be given a United Nations title.

A study published Nov 2021[9] by E. Martinez of MIT, Cambridge MA and C. Winter of Istituto Tecnologico Autonomo de Mexico indicated that just over thirty percent of 1016 US residents surveyed supported the granting of personhood status to AI. This support was given on the basis of whether or not it could be determined that the AI had the ability to feel pleasure or pain. The percentage varies considerably with political ideology, with substantially different patterns among liberals and conservatives. Pew Research Center's 2023–24 Religious Landscape Study [10]found that among self-described liberals, only 37 percent identify as Christian, while 51 percent are religiously unaffiliated. About 11 percent attend religious services at least once a week, and only 14 percent say that belief in God is necessary in order to be moral and have good values.

Our modern age has seen millions become addicted to screens, spending hours and days on video games or social media group chats, sometimes to the detriment of their real-world relationships or even their health. But what happens when the screen is inside your head? A life enhanced by BCI might be dazzling, but what if the connection is lost? Beyond feeling 'out of touch', we risk becoming so intertwined with technology that its absence could make us feel incomplete, or worse, incapacitated. In essence, we would become slaves to the very tech designed to liberate us.

Public attitudes are shifting as well. Some surveys suggest that a substantial minority of people are open to granting some form of personhood to AI if it appears capable of pleasure, pain, or emotional experience. That shift reflects a culture increasingly willing to interpret simulation as reality, and performance as identity. Once that concept is established, the line between human dignity and artificial imitation becomes easier to confuse.

Meanwhile, ordinary technology has already shown how easily dependence can take hold. Millions of people now struggle to disconnect from screens, social platforms, and constant streams of digital input. These technologies shape attention, desire, mood, and relationships. If that level of influence is possible through devices we hold in our hands, then neural technologies raise the stakes considerably. What happens when the interface is no longer in front of the face, but within the body itself?

Dependence would likely deepen in several ways. A person enhanced by BCI systems might gain speed, access, convenience, or even social advantage. However, such gains could create a new kind of vulnerability. If the system fails, is removed, or becomes inaccessible, the result may not feel like inconvenience. It may feel like loss of function, loss of identity, or loss of control. A tool that once promised freedom could create a new form of bondage.

There is also a social dimension. Nearly every major technology begins with limited access. At first, only the wealthy or well-positioned can afford it. If BCIs move beyond therapy into enhancement, the result could be a new form of inequality rooted not only in money, but in cognition, performance, and opportunity. Enhanced individuals might gain educational, professional, or social advantages unavailable to others. This would widen economic gaps, creating a deeper division in the structure of society.

That possibility raises additional political and ethical questions. If enhancement becomes culturally normal, will governments or institutions eventually frame it as a right? Will public systems be pressured to subsidize access? If so, who will bear the ongoing cost of maintenance, connectivity, updates, and security? And, who will govern the systems to which enhanced individuals remain permanently linked? Any network that supplies information can also shape what is considered relevant, urgent, acceptable, or true.

The mental and emotional implications are equally serious. Human consciousness is not a blank platform onto which new capabilities can be added without consequence. Our minds are ordered through memory, emotion, habit, judgment, and relationship. Introducing AI-driven interaction into that structure may produce effects no one can fully predict. Information overload, confusion of identity, altered emotional processing, and new forms of psychological instability are all plausible concerns. Even so-called therapeutic uses could create new disparities between those who can afford AI-mediated mental support and those who cannot.

Standards would become another point of concern. If neural enhancement develops at scale, someone will decide what counts as normal, improved, optimized, or desirable. Those standards will reflect the priorities of institutions, corporations, developers, and governments. Bias, profit motives, and political interests would not remain outside the system. They could become part of the system itself.

**Deep Dive Box: Human augmentation and faith.**

Technology can clearly do remarkable things. The issue, however, is whether every increase in power should be accepted as progress. Scripture teaches that human beings are not raw material for endless redesign. We are created beings, made in the image of God, with dignity that does not depend on performance, efficiency, or enhancement. A worldview shaped by biblical truth cannot treat the human person as a platform for technical upgrades without asking what is being lost in the process.

We should all be grateful for tools that relieve suffering and restore function. Compassion should welcome healing. But gratitude for healing is not the same as surrender to a vision of humanity that seeks transcendence through machinery. We must learn to distinguish between care for the body and attempts to redefine the person.

As Christians, we are called to understand the world clearly, and to live truthfully within it, not ignore it. That means learning enough about these technologies to recognize both their promise and their danger. It also means forming judgments through Scripture rather than through marketing, celebrity influence, or cultural excitement.

We should think carefully before accepting systems that move human beings toward deep integration with AI-driven networks. Furthermore, we should ask what the companies behind these technologies assume about the body, the mind, freedom, identity, and the soul. We should look for pastors and leaders who show discernment, moral seriousness, and theological clarity. And we should resist the impulse to hand over spiritual care to automated systems, whether in the form of sermon generation, therapeutic bots, or digital imitations of Christ.

The frontier of AI-enhanced human augmentation is moral and spiritual. We do not serve the world well by ignoring it, and we do not serve it well by condoning every innovation as progress. While the world moves quickly in another direction, we must hold fast to a Biblical vision of the human person.

### The Bright Side: BCIs could help those with disabilities communicate or move

BCIs might even pave the way for enhanced learning experiences. Brain-computer interfaces could become one of the most compassionate uses of advanced technology when directed toward restoration rather than enhancement. For people who are mentally alert but trapped by paralysis, ALS, spinal cord injury, or severe motor impairment, the ability to move a cursor, type a message, control a device, or communicate a basic need through thought-directed systems could be life-changing.

This is not a small benefit. Communication is deeply tied to relationship, dignity, care, and participation in ordinary life. A technology that helps a person speak when the body can no longer cooperate deserves careful attention and gratitude. From a Christian point of view, tools that relieve suffering and restore function can reflect mercy when they serve the person rather than redefine the person. The moral clarity comes from the purpose: helping the vulnerable live more fully within their humanity.

⚠ **Proceed with Caution: The brain is our final frontier.**

Fiddling with it means delving into the essence of our consciousness, emotions, and memories. The danger begins when a healing tool becomes a doorway for deeper integration with systems that can monitor, shape, influence, or commercialize human thought. A brain-computer interface is not merely another screen. It reaches into the body and touches the neural processes connected to movement, perception, intention, and possibly emotion or memory. That means the usual concerns about data privacy, cybersecurity, corporate control, and addiction become far more serious. A hacked account can be replaced. A compromised neural interface raises questions about agency, identity, and consent. Christians should be especially alert when medical language is used to normalize enhancement, when restoration becomes a bridge to redesign, or when human limitation is treated as a defect to be erased. The central question is not whether the technology is impressive. It is whether the person remains honored as an image-bearer of God, or becomes a platform for upgrades, surveillance, and control.

## 4.4 POST HUMAN WORLD

**In short:** What happens when humans are unnecessary? The following narrations is based on actual events.

The train was full. The man found his seat, stowed his bag, opened his phone. The message shocked him to his core. A government notification, three lines long, informing him that his social credit score had fallen below the threshold required for high-speed rail travel. The ticket he had purchased was valid. The seat was his. But the system had updated overnight, and the update had reached the station before he did.

He was removed at the next stop. This is not a hypothetical. Human Rights Watch has documented cases of individuals blocked from travel without notification of the specific infraction or avenue for appeal (HRW, 2019)[1].

Lawyer Li Xiaolin was on a work trip in 2016 when he tried to buy a plane ticket home to Beijing. The online system rejected his national identity card. He checked the court's website and found his name on a list of "untrustworthy" people. This was the result of a 2015 court order he believed he had fulfilled. He was stranded more than 1,200 miles from home. It took three weeks before an official spoke to him. The court had deemed his written apology "insincere."

China's AI surveillance infrastructure tracks citizens across financial transactions, online behavior, and physical movement, with software alerting officials automatically when a violation is detected. The system integrates financial records, social interactions, and legal infractions into a single score. The knowledge that one's actions are being monitored has measurably increased self-censorship and conformity to state-defined norms.

The man on the platform had not committed a crime. But his behavior had shifted from the statistically permitted range. Such algorithms might dock points for a late bill payment without knowing the person was in hospital. Or penalize a business for filing taxes late without knowing they were recovering from a flood. The context in which events take place is expensive to provide to the AI. Scoring is cheap. The system does not distinguish between them.

Research has found the system particularly disadvantages rural residents, reinforcing structural inequalities rather than correcting them. The people with fewest resources to navigate bureaucratic error are the people most exposed to it, and in some ways that makes it harder to oppose. Dystopia is easy to recognize. A filing error that costs you a train seat is not.

The EU's AI Act of 2024 explicitly bans AI systems [2]involving social scoring by public authorities, citing threats to democracy, freedom of expression, and human dignity. The prohibition exists because the risk is real.

The man on the platform eventually reached his destination. However, it took four hours longer than necessary. No one reviewed his case because there was no case. There was only a number, and the number was wrong, and the system had no procedure for that. With these considerations in mind, can autonomous machines still be considered valuable assets to humans, or replace certain human roles entirely? This growing concern is not limited to individual jobs, but extends to how societies organize work.

What happens if humans become economically or functionally unnecessary in key areas of society? Technology has already displaced certain forms of manual or intellectual labor. Automation replaced manual processes in manufacturing. Software reduced the need for clerical work. Each wave of change shifted employment in a new direction, rather than eliminating it entirely. New roles emerged as others disappeared. As of the writing of these lines, by 2030 economists estimate that most labor force could be replaced by AI. This change doesn't just affect intellectually rich jobs in law, healthcare or computer science. This refers to all jobs across the markets in developed societies. As AI capability expands, work that once required education and experience may become partially or fully automated.

This creates the possibility of much wider **economic displacement.** If machines perform a greater variety of tasks more efficiently, demand for human labor will decline in more and more sectors. Productivity may increase, but the distribution of opportunity will change. Some individuals may adapt by moving into new roles. Others may find fewer options available.

The issue goes beyond the financial implications of employment. Work has long shaped identity and participation in society. When decision-making or production shifts to machines, the structure of human contribution to society changes from producing value to managing systems that produce value. The transition may be gradual, but it alters the relationship between physical or intellectual work and outcome.

Therefore, if productivity depends primarily on automated systems, ownership of those systems becomes significant. Those who control the infrastructure gain unprecedented control over output and distribution. The concentration of technological capability can therefore translate into concentration of economic power.

This leads to the second concern: **political power shift**.

Technology has always influenced governance, but AI adds another layer. Predictive models managed by computers assist with planning, resource allocation, and risk assessment. In some cases, governments rely on these tools to guide policy decisions. Specialists design the systems involved in the process. Administrators apply them. However, it is the citizens who experience the outcomes. The chain of decision-making grows longer, pushing its operators far apart from each other. Accountability for a particular action can become less direct, and eventually fade from a particular individual, to a group, and eventually to no-one in particular.

If AI systems become central to planning and administration, those who design and maintain them gain influence and control over the political infrastructure. Decisions that are essentially technical, rather than political, end up shaping social outcomes.

This dynamic can shift how power is exercised at the highest levels. Political governance may increasingly rely on optimization of processes and prediction of outcomes, using efficiency as a guiding principle at least on the surface. However, political and social life involve values that are not purely technical or quantifiable. Fairness, representation, and dignity cannot be calculated in the same terms as optimization.

The third concern follows naturally from these realizations: **technological control over populations**.

Artificial intelligence systems require centralized data, computing resources, and continuous refinement. Control over this infrastructure influences how systems operate, as well as affecting what objectives they pursue. When critical decisions rely on AI, access to the hardware and technological systems used to manage it becomes important. Those who control them influence how information flows, how resources are allocated, and how risks are evaluated.

This raises questions about **autonomy.** Individuals increasingly rely on systems for navigation, communication, and critical decision support in economy, social and military applications. Institutions often depend on AI for economic planning, forecasting, and analysis. As world-wide reliance on AI increases, overall independence of society from the systems once used as tools may decrease. Technology, therefore, becomes integral part of the very same environment which it is tasked to analyze, and choices are made on the basis of potentially skewed data.

The idea of human obsolescence does not imply that workers will disappear at all levels of industry. If machines handle analysis, planning, and production, human participation is forced to change. The issue, however, is *how* society responds when that happens. Economic participation, political authority, and technological control are all critically dependent on how that adjustment occurs. Intelligence has long been associated with human identity, often placed at the same level as the human essence. As machines perform more cognitive functions, however, that association becomes less clear, redefined in ways that are less dependent on human factors. Technology is reshaping our world, and AI is the closest tool to human intelligence that has ever been crafted. The final outcome on how we control it and use it to improve human condition will depend on how this transition into the era of intelligence is understood and guided.

**Deep Dive Box: The shift from wages to infrastructure**

The deepest issue may not be job loss by itself. Societies have absorbed labor shocks before. The sharper change comes when economic value moves away from human effort and toward technical infrastructure. The work of the Organization for Economic Co-operation and Development (OECD) on AI and regional labor exposure[3] points in that direction. The pressure will not fall evenly. Some jobs will be redesigned, some reduced, and some displaced, but the larger consequence is structural. The more value depends on models, compute, data, and distribution platforms, the more power gathers around whoever owns those systems. In that world, the central question is no longer, "Who has work?" It becomes, "Who owns the machinery through which work, access, and opportunity are organized?" That is a deeper form of dependence than unemployment alone.

### The Bright Side: The human question

The more AI forces society to confront labor, power, and control, the harder it becomes to talk only about efficiency. Institutions are having to recover older questions that modern systems often try to avoid. What kind of dependence is acceptable? Which decisions must remain contestable because they bear on dignity, not just output? OECD's recent work[4] on human-centered adoption of trustworthy AI in the workplace, and the National Institute of Standards and Technology's ( NIST) risk-management efforts both reflect a turn toward accountability, oversight, and human responsibility.

### Proceed with Caution: A society can remain busy while becoming less free

When production, planning, and distribution are routed through systems that few people understand and even fewer people control, dependence expands in subtle ways. Services still function. Markets still move. Benefits still arrive. Yet the terms of participation have changed. Citizens become users of systems they did not shape. Workers become adjuncts to platforms they do not govern. Everyone keeps participating in society while losing their role within it.

## 4.5 CRISPR AND AI: GENETIC SCISSORS TO EDIT LIFE

**In short:** CRISPR, IVF, PGD, and AI give us growing power to edit, rank, and in some cases even select human life, but that power can quickly shift focus from healing to control. For Christians, the central concern should be to preserve our view of human life as a gift from God rather than let it be turned into a managed product of design. The following narration is loosely based on the use of AI in reproductive health applications.

The clinic had photographed each of the five embryos at regular intervals, tracking the rate of division, the symmetry of cells, the texture of the outer shell that would dissolve at implantation. A scoring system translated these observations into grades. The grades were attached to probabilities, crafted into a recommendation that came in a letter.

One embryo, designated E2, had been assessed by the clinic's AI-assisted selection system as the strongest candidate for transfer. The system had reviewed the images, Jane's age, and the clinical data from previous cycles. It had cross-referenced these against outcomes from thousands of prior transfers. It had assigned E2 the highest likelihood of resulting in a normal pregnancy.

Charles and Jane read the letter at a kitchen table in the evening.

They had not been told which embryos were graded lower, or what the system predicted for E3, E4, or E5. They had been told that E2 was the recommended transfer. Proceeding with a different embryo would require a conversation with their physician.

At the follow-up appointment, Jane asked what would happen to the others. The answer, delivered gently, was that many options were available. Unused embryos could be frozen indefinitely, donated to research, donated to another couple, or allowed to cease development. The decision was theirs. The clinic was there to support whatever they chose.

What the clinic did not say was how it had weighted the variables. It had not been designed that way. They didn't disclose on what basis the algorithm had determined which mattered more: the image at hour 68, or the maternal age? What margin separated E2 from E3? Was the difference between them the difference between a child and the absence of one?

The parents transferred E2. The pregnancy was successful.

They still think about the other four.

Editing the code of life is one of the clearest expressions of the transhumanist project, morphing a long-standing human desire into a practical ambition with a clear focus. Today, many powerful and intelligent people don't just seek to understand life. They want to direct, revise, and shape it in all its future form.

Few scientific developments have captured modern imagination as powerfully as CRISPR.[1] The technology is often described as a breakthrough in gene editing, and that description is justified. CRISPR, short for "clustered regularly interspaced short palindromic repeats," is a gene-editing technology adapted from a natural defense system against viruses found in bacteria. This approach made genome editing faster, cheaper, and more precise than earlier methods, which is why CRISPR quickly became one of the most important tools in modern biotechnology. CRISPR gave scientists a way to identify, cut, and alter specific sections of genetic material with a precision earlier methods could not match. Faulty sequences may be removed, altered, or replaced. Genes may be switched on, or off, or modified in ways that reach deeply into the biological structure of living things.

Genetic diseases that once seemed untouchable may become treatable. New therapies may emerge for conditions that have long resisted intervention. Crops may be altered to withstand drought, disease, or environmental stress. These prospects help explain why CRISPR is often presented as one of the most important biotechnologies of the modern age.

Yet the power to rewrite genetic code is not a small matter. The question is not only what can be done, but what should be done, and on what grounds. A tool this powerful requires moral restraint. Once human beings gain the ability not only to read the code of life but to revise it, biology enters a landscape of morality and ethics.

Part of CRISPR's influence comes from its accessibility. Earlier forms of genetic modification were often expensive, technically demanding, and difficult to scale. CRISPR lowered those barriers to the financial access of most entry-level companies. It made gene editing more practical for laboratories across the world, and accelerated the pace of experimentation. That matters because a technology that becomes easier to use also becomes harder to contain. Ethical uncertainty moves quickly into reality and practice..

CRISPR itself was first described in 1985 in the bacterium *Escherichia coli* by Ishino and colleagues. Its development has been marked by controversy. Patent disputes, competing claims of discovery, and debates over recognition followed almost immediately. In 2020, the Nobel Prize in Chemistry was awarded to Jennifer Doudna and Emmanuelle Charpentier for their work in developing CRISPR-Cas9 as a method for genome editing. That award reflected the scale of the scientific achievement. It did not settle the deeper questions now attached to it. By 2015, the technology had already been used on human embryos, without specific oversight except for local, institutional review boards.[2] In only a few years, CRISPR moved from basic discovery to one of the most consequential tools in modern genetics.

Researchers have explored the use of CRISPR in conditions such as sickle cell disease, cystic fibrosis, and some cancers. In these settings, gene editing is often framed as repair. A harmful mutation may be corrected. A dysfunctional process may be interrupted. A serious inherited disorder may one day be treated more effectively than before. Christians should not treat the relief of suffering as morally irrelevant, but we must understand the landscape in which modern science is working. Medicine is one way we exercise stewardship and compassion in a fallen world, but the same technology that may repair disease can is currently being directed toward enhancement, selection, and control. Once gene editing moves beyond therapy and into the design of preferred traits, the moral framework changes. The issue is no longer healing what is damaged. It becomes the shaping of human beings according to desired outcomes. At that point, medicine begins to merge with engineering, and the human person becomes a project with optimal targets and deadlines.

History prompts us to be cautious. Genetic intervention has already produced serious harm. In 1999, Jesse Gelsinger died[3] after complications related to gene therapy of his metabolic disease. His death became a warning about the cost of moving too quickly in a field where biological systems are only partly understood. The field of bioethics exists *because* scientific and medical authorities have repeatedly, and sometimes purposefully, failed to restrain themselves. The Nuremberg Code, the Declaration of Helsinki, and the Belmont Report were all written in response to human abuses carried out in the name of research, expertise, or progress. There is no single worldwide code for bioengineering that functions exactly like them. Instead, human genetic engineering is governed by a layered set of international declarations, professional reports, national laws, and institutional review systems. The closest global ethical frameworks include UNESCO's Universal Declaration on the Human Genome and Human Rights, UNESCO's Universal Declaration on Bioethics and Human Rights, and the World Health Organization's 2021 recommendations and governance framework on human genome editing. Together, these documents emphasize human dignity, informed consent, nondiscrimination, transparency, public accountability, protection of vulnerable persons, and caution regarding heritable genome editing. Their influence is significant, but their legal force depends on how individual nations incorporate these principles into law and regulation.

UNESCO frames the human genome[4] in relation to human dignity, human rights, nondiscrimination, informed consent, privacy, and the responsibility to ensure that advances in genetics serve humanity. UNESCO describes the human genome as part of the "heritage of humanity," and the declaration states that benefits from advances in biology, genetics, and medicine should be made available with respect for each person's dignity and rights.

The World Health Organization's (WHO) 2021 reports[5] are the closest current global governance framework specifically aimed at human genome editing. The WHO Expert Advisory Committee developed both recommendations and a governance framework for oversight of human genome editing. The WHO materials address clinical trials, registries, transparency, whistleblowing, international collaboration, equity, safety, and responsible oversight.

Expertise does not guarantee moral clarity.

Former National Institute of Health (NIH) director and leader of the Human Genome Project Francis Collins, a geneticist and evangelical Christian, has argued that genetic medicine is compatible with faith[6(116)] and that the image of God is not located in our DNA. Christian faith and genetic medicine need not be opposed. In his account of human uniqueness, the image of God is tied to God's gift of "free will" and "a soul," rather than to DNA itself. Collins therefore supports therapeutic uses of genetic medicine while warning against embryo and germline interventions. However, the dividing line between genetic intervention and modification is increasingly blurred. The advances in the field of epigenetic have clarified we know far less about gene expression than we thought. The distant effects of a deceptively simple gene modification are not fully predictable, and neither is the possibility of downline transmission.

Ethical frameworks for germline genetic engineering remain limited and uneven. This has enormous implications, since this technology targets reproductive cells, affecting humanity as a whole. Germline modifications do not simply affect one individual, but reaches into future generations. Changes introduced at that level may be inherited by descendants who had no voice in the decision and no ability to escape its consequences. This invasive action could endanger the biological continuity of human life.

The best-known example of unregulated reckless behavior in genetic modifications came in 2018. As we previously discussed, the Chinese biophysicist He Jiankui announced the birth of twin girls whose embryos had been genetically modified using CRISPR-Cas9. The edits were intended to alter the CCR5 gene in a way that might reduce vulnerability to HIV infection.[7] The announcement provoked global outrage, not only because the procedure crossed a widely recognized ethical line, but because the long-term consequences were unknown. Off-target mutations, unpredictable developmental effects, and inherited biological consequences had not been adequately understood. A third genetically edited child was later reported. The scandal that followed led to criminal penalties, but it also proved something larger. This technology has already outrun the ethical structures meant to govern it.

Some nations and institutions have expressed concern, drafted recommendations, or restricted particular forms of experimentation. Yet, the global landscape remains fragmented. Public funding may be restricted in some countries while private funding continues to push research and experimentation forward at breakneck speed with limited oversight. International standards remain weak, uneven, or difficult to enforce. This patchwork approach leaves decisions with profound consequences vulnerable to political pressure, private ambition, and international competition not limited to medicine. Discussions about biological enhancement have increasingly entered military and geopolitical contexts. Once a technology promises power, nations will have strong incentives to pursue it, especially when rivals appear willing to do the same. Even when particular claims are exaggerated, the underlying pattern is clear. Human biology is increasingly viewed not as a gift to be treasured and respected, but as raw material to be optimized. A starting point. A good initial version of something far greater that the engineers envision for our future.

Once humans are designed for advantage, whether medical, social, or military, the person is reduced to function. Identity becomes subordinate to utility. The body becomes something to engineer toward a target. That moral shift changes the meaning of the human being from someone with dignity to someone assessed by performance.

The most dramatic consequences of this reckless behavior may soon manifest plainly for all to witness, but that point may be already past the stage of irrecoverable damage. Humanity is targeting the code of life with the presumption of understanding its secret at a level that we simply do not have. We treat traits and genes as if they were written by us in computer code, a woefully inadequate rendition of the complexity embedded in our genome. Genes do not operate in isolation. They interact within systems that are layered, with dynamic processes controlled by operating functions with multiple variables such as time, chemistry, nerve and hormonal inputs, and still only partly understood. A desired change may produce unexpected effects elsewhere. Off-target mutations, abnormal development, long-term health complications, and inherited instability remain legitimate concerns. When those risks are placed upon embryos, the problem becomes more dramatic, because the subject of the intervention cannot consent, and cannot later reverse what has been done.

Once humanity recognizes that embryos are human lives at an early stage of development, the ethical discussion changes completely. Embryos are not laboratory material to be used, altered, ranked, or discarded according to technical goals. They are sons, daughters, brothers, sisters, neighbors. They are image-bearers. They are persons with moral worth, even before birth. That conviction places a far heavier burden on any proposed use of biotechnology at the embryonic stage.

This is where IVF, embryo screening, and AI-driven selection converge into an acutely pressing point. In reproductive medicine, AI is now being used to assist with embryo selection, improve predictions about viability, and refine decisions that were once made primarily by human observation. At one level, the appeal is obvious. If technology can help identify embryos more likely to survive implantation, or reduce the likelihood of passing on severe genetic disorders, many will see that as compassionate and practical. In a setting where disappointment and loss are common, greater precision naturally sounds like progress.

But the moral questions do not disappear because the process becomes more accurate. As reproductive science becomes more dependent on algorithmic systems, it also becomes easier to view embryos in terms of ranking, desirability, and optimization. Human life at its earliest stage is no longer merely received. It is assessed, compared, filtered, and selected. What begins as an effort to improve outcomes can gradually train a culture to think of children not as gifts, but as managed results. Human beings *are not* products to be refined until they meet a preferred standard. The more reproductive technology moves in the direction of selection and design, the more likely it becomes that difference, limitation, and unpredictability will be treated as defects to eliminate. A given skin or eye color may be preferred to another. A certain athletic ability chosen for its advantages. The list is endless.

From a Christian perspective, this is a decisive issue. Human dignity does not begin when a child becomes wanted, healthy, socially desirable, or genetically promising. It does not rest on projected intelligence, physical normality, or freedom from disease. Human life has value because it is given by God and bears His image from the beginning. That conviction must govern how we think about reproductive technology.

AI also extends its control within IVF by identifying patterns associated with successful implantation and development.[8] Instead of relying only on the trained judgment of embryologists, clinics can now use machine learning models, neural networks, and computer vision systems to evaluate embryos with greater speed and consistency. Time-lapse imaging allows continuous monitoring of embryonic development. Algorithms analyze cell division, timing, morphology, and other markers that may correlate with viability. AI systems may also assist in interpreting genetic screening data, offering increasingly detailed estimates of chromosomal normality and developmental potential.

This is often presented as an advance in objectivity. Human observers may disagree or miss details. An algorithm can review thousands of cases, detect patterns across large datasets, and apply those patterns consistently. Equitably. In that sense, AI can standardize a process once shaped more heavily by personal judgment. Even so, objectivity in method does not remove the moral weight of the decision. A system may become more accurate at selecting embryos, but do so without answering whether embryos should be treated as selectable material in the first place. Better prediction is still being used in service of a prior act of evaluation. The ethical problem is not solved by improving the tool.

This is why the language of viability must be handled carefully. In clinical use, viability refers to the likelihood that an embryo will implant or develop successfully. That may be a useful medical category. But it can also encourage an attitude in which worth becomes tied to projected outcome. Christian ethics cannot accept that shift. The value of a human life is not measured by its likelihood of success, physical normality, or developmental promise as judged by a system.

Pre-implantation genetic diagnosis (PGD) adds another layer to this problem.[9] PGD allows embryos created through IVF to be tested for specific genetic conditions before implantation. It is used to identify embryos that carry known mutations, chromosomal abnormalities, or inherited disorders. For parents with a family history of serious illness, this can sound responsible and compassionate. But PGD does more than provide information. It changes the meaning of selection. Embryos are no longer assessed only for visible signs of development or likely implantation success. They are also examined for the traits they carry. The choice of which embryo to implant can then be shaped not only by survival potential, but by genetic preference. At that point, the process moves further away from treatment and further into judgment.

This distinction is crucial. PGD does not cure an embryo. It does not heal a child in the womb. It identifies some embryos as genetically preferable to others, and that information is then used to decide which embryos will be given the chance to continue developing. The issue is that human beings at the embryonic stage are being sorted according to genetic criteria before birth. Supporters of PGD often frame it as a way to prevent suffering. In many cases, that concern is sincere. Parents may fear passing on devastating illnesses. Physicians may want to reduce the burden of disease. Those motives should be taken seriously. We should all care about reducing suffering in the world. But the moral question remains. Does the prevention of suffering justify the creation, testing, selection, and likely disposal of some embryonic human lives in order to favor others?

From a Christian perspective, the answer cannot rest on outcomes alone. An embryo is not more worthy of life because it is free of a certain condition, and not less worthy because it carries one. Human dignity does not rest on health, predicted ability, or biological normality. It rests on the fact that every human being bears the image of God.

This is where AI may change the future of PGD in ways that are easy to underestimate. Today, AI helps analyze large datasets, improve pattern recognition, and rank embryos according to selected variables. Tomorrow, more autonomous systems could move further. They may not merely assist a physician. They may recommend which embryos to implant, freeze, or to discard, and which traits to prioritize. This would be done using decision trees too complex for parents or clinicians to follow in detail. The more these systems gain authority, the more the logic of selection can become detached from direct human moral reasoning.

That prospect matters because **autonomous AI does not need malice to do harm**. It only needs optimization goals. A system trained to maximize implantation success, minimize projected disease burden, reduce cost, or improve preferred outcomes will sort human embryos according to those metrics with relentless consistency. It will do exactly what it was built to do. If the categories are wrong, the scale of the wrong may become enormous. In that environment, the danger is not only that human beings will make eugenic choices; it is that those choices will be normalized, automated, and concealed behind the language of efficiency, objectivity, and clinical decision support. Parents may be told they are simply following the best recommendation. Clinics may defer to systems too complex to explain clearly. Institutions may begin to treat algorithmic selection as standard care. Once that happens, moral responsibility becomes easier to diffuse even as moral damage expands.

The categories may also broaden quickly. Selection may begin with grave diseases, then expand toward lower-risk traits, nonfatal disorders, sex selection, projected cognitive ability, behavioral traits, or preferred probabilities tied to future performance. As we all know, and have witnessed in our society, the technical line can move faster than the moral one. A culture that becomes comfortable screening embryos for one reason may eventually become comfortable screening them for many. Autonomous AI could accelerate that transition by making the process faster, cheaper, more comprehensive, and more difficult to question in its appearance of equity.

This reflects a larger cultural pattern in which modern technology promises more control, more certainty, and fewer unwanted outcomes. Human life is uncertain, and uncertainty is hard to bear. But the desire to reduce uncertainty can become the desire to eliminate dependence, surprise, and limitation altogether. When that logic enters reproduction, it changes how children are imagined. The child is no longer simply received, but managed in advance.

Not every use of technology in fertility care is morally

identical. Medicine can be used to support the body, treat disease, and assist couples in ways that respect life rather than dominate it. But there is a real difference between helping the body and subjecting nascent human life to increasingly refined systems of selection and control. AI does not erase that distinction. It sharpens it.

The combination of CRISPR, IVF, PGD, and increasingly autonomous AI therefore raises one central question. Are children, and human beings, gifts to be received with gratitude, or products to be optimized through increasingly precise forms of selection? That question reaches beyond medicine into the moral imagination of an entire civilization.

As Christians, we need clarity. The image of God in human beings places limits on what may rightly be done, even in the name of compassion, efficiency, or scientific progress. Any technology that encourages us to sort, rank, redesign, or discard human life at its beginning must be examined with the greatest seriousness.

CRISPR, on the other hand, places enormous power in human hands that are not necessarily able to govern it. AI amplifies that power, and PGD channels it toward selection. None of these tools can tell us what a human embryo is worth. That answer does not come from a dataset, a neural network, or a success-rate chart. It comes from a prior truth. Human life is not ours to value according to preference. It belongs first to God.

A culture that learns to edit life will soon try to score it. A culture that learns to score life will begin to sort it. When that happens, technology is no longer serving humanity, but steering it toward a target outcome that it chose as optimal on its own accord. The Christian response to this attack must begin there. We do not deny that these tools hold real power, or that some uses may relieve genuine suffering. But we must refuse the logic that turns children and human beings into ranked outcomes, and human worth into a technical judgment. Human life is a gift, not a project. If we lose that truth, no amount of intelligence, artificial or otherwise, will recover what we have optimize ourselves to discard.

The *Imago Dei* is not a property of biology. It cannot be upgraded by inserting a tardigrade gene, and it was not diminished in Lulu and Nana by the edit He Jiankui made. But the idea behind these experiments, that the human being is a problem to be solved, a platform to be optimized, is a challenge that Scripture addresses directly. It is as old as the tower of Babel and the fall of the Enemy. It dresses differently in every century, but its logic is the same: that what God made is not enough, and that human hands can do better.

The question germline editing puts before the church is not primarily technical. It is theological. Who is the author of human life? And what does it mean to rewrite what God has written?

Those questions are the right ones to ask before the next announcement, and the next pair of children born into a world that did not ask their permission.

**Deep Dive Box: Lulu and Nana**

He Jankui was sentenced to three years in a Chinese prison, but the condemnation focused mostly on procedure — flawed consent, hidden research, regulatory violations. It largely avoided the harder question. It focused on how He did what he did, in absence of regulatory oversight, and keeping the world academia in the dark about his pursuit. But what he did, and what it means for the children who now carry those edits in every cell of their bodies, and who may one day pass them to children of their own, never quite reached the level of concern[10].

Unlike gene therapies that alter somatic cells (the cells that make up the non-reproductive part of the body) as a medical treatment, the editing performed on Lulu and Nana made a permanent change to the germline : a change that could be inherited by future generations.[(92)] That is the line He crossed. Not merely the line of scientific consensus, but the line that separates treating a person from rewriting one.

We should think carefully about where that line is, and why it matters.

Scripture teaches that every human being is made in the image of God — the *Imago Dei*. Genesis 1:27 does not say that some humans bear this image more fully than others, or that the image can be improved upon by biotechnology. It says that God made humanity in his image, male and female, and called it very good. That declaration is not conditional on a particular trait, ability, or genetic sequence. It does not depend on resistance to disease, or survival of radiation. It is an act of divine authorship, and it applies to every person who has ever lived, including two girls in China whose names we do not know, who did not consent to what was written into them before they were born.

It applies to the entire human race, which did not consent to being edited by some rogue scientist playing God in an unauthorized laboratory.

**Deep Dive Box: The mythical super soldier**

In March 2023, a team of military medical scientists in China reported that they had inserted a gene from the tardigrade (a microscopic animal known for its extraordinary resilience to environmental damage, including radiation) into human embryonic stem cells, and had significantly increased those cells' resistance to radiation. The paper came from the Academy of Military Sciences in Beijing [11]. Researchers suggested the work could lead to soldiers capable of surviving nuclear fallout. This was not science fiction. It was a published study, funded by a military institution, aimed at producing human beings with capabilities engineered for the battlefield.

The logic here is worth naming clearly. If a soldier can be made radiation-resistant by editing germline cells, those changes persist, and they are passed on to children. The engineered trait enters the human population, and does not leave. What begins as a military program becomes a heritable condition of human existence, decided not by the people who carry it, but by the states and institutions that ordered the research.

From a Christian standpoint, this is not enhancement. It is replacement. An erasure of the boundary between creature and artifact.

### The Bright Side: An hopeful solution for sickle cell disease

Not every application of gene editing raises these concerns. In December 2023, the U.S. Food and Drug Administration approved Casgevy, the first CRISPR-based therapy cleared for clinical use in the United States, as a treatment for sickle cell disease.[12] The therapy works by editing a patient's own blood stem cells outside the body. It corrects the defect that causes the disease, and those cells are then infused into the patient. Nothing is written into the germline. Nothing passes to the next generation. The changes are contained to the person being treated, and the goal is straightforwardly to relieve suffering. This is gene editing in the service of healing, not redesign. It respects the boundary between medicine and manufacture.

### Proceed with Caution: Humans as products

He Jiankui's attack on the human genome did not manifest in a vacuum. It arrived at the end of a pipeline that had been building for decades. In vitro fertilization, once deeply controversial, is now routine. Preimplantation genetic diagnosis (PGD)  is widely accepted in medicine as a way to prevent heritable disease. PGD is the practice of screening embryos created through IVF, and selecting which ones to implant, discarding or selling the rest, depending the particular institution's policies. The logic of each step seemed reasonable on its own terms. But the combined effect of IVF and PGD is that human embryos are now regularly created, evaluated, and discarded based on their genetic makeup[13]. CRISPR does not introduce a new kind of decision into that pipeline. It extends a decision already being made, extending it from the process of selecting among embryos to rewriting them. Christians who are comfortable with IVF and PGD, and troubled by germline editing, should ask whether their comfort with the former has made the latter easier to arrive at than it should have been. The question is not only where the line is now. It is how the line has moved, and where does it move next.

# PART FIVE

## *The Christian Answer*

How does our understanding of human nature shape our response to technology? The Christian tradition approaches this question by placing God at its center, and His definition of who we are as the fundamental litmus test to evaluate any emerging technology. Human identity cannot be defined solely by biology, intelligence, or productivity. It is grounded in our relationship with the Creator. We are image-bearers of God, possessing dignity that does not depend on our personal capability. Creativity, moral awareness, and self-reflection are not accidental features. They reflect a deeper origin.

The goal therefore should not be to simply extend human capability, or overcome anatomical limits, but to steward human life responsibly. AI tools can assist, heal, and support, but should be measured by how they serve the human person, rather than by how they replace it, or redefine it. Our emphasis should shift from the aspiration to become more than human, to understanding *what it means* to be human. And the answer to that question is rooted in God's message to us through the Bible.

## 5.1 INTELLIGENCE AND THE CREATOR

**In short:** The appearance of intelligence does not imply its presence.

The discussion of artificial intelligence repeatedly returns to a basic question. What is intelligence, and where does it come from? We can describe how intelligence behaves. We can model certain aspects of reasoning. We can build systems that simulate decision-making. Yet the origin of intelligence itself remains difficult to explain.

Intelligence is not only the ability to process information. It involves understanding, reflection, creativity, and the search for meaning. Human intelligence asks questions about truth. It evaluates ideas, recognizes beauty, and considers moral responsibility. These qualities suggest more than mechanical processing. They point toward awareness that interprets rather than merely computes.

The presence of intelligence invites explanation. When we encounter structured information, we often infer intention. A written message implies an author. A coded sequence suggests a source. The inference is not arbitrary. Order that carries meaning typically arises from a mind. This pattern shapes how humans interpret the world. When we recognize intelligence, we look for its origin.

Theological reflection approaches this question directly. In Christian thought, intelligence is not accidental. It reflects the character of a Creator who is understood as rational, purposeful, and relational. Human intelligence is seen as a reflection of that source. The ability to reason, create, and seek truth is understood as part of being made in the image of God.

This perspective does not deny biological processes. The brain remains essential. Neural activity supports thought. Memory depends on physical structure. Yet the presence of awareness suggests that *intelligence is more than the arrangement of matter.* The mind interprets information and recognizes meaning. This capacity points beyond the mechanisms that support it.

The argument can be stated simply. **Intelligence implies design.** When intelligence appears, it suggests an origin that is itself intelligent. The reasoning does not claim that every detail of intelligence is understood. It recognizes that awareness, meaning, and purpose are not easily reduced to random processes. The existence of intelligence becomes a clue to its source.

This idea contrasts with a purely material explanation. If intelligence emerges solely from physical interactions, then awareness must arise from complexity alone. Matter arranged in sufficient detail produces thought. The challenge lies in explaining how this transition occurs. Complexity can describe structure, but it does not necessarily explain experience. The step from processing to understanding remains difficult to define.

The theological perspective offers a different framework. Intelligence originates from a mind rather than emerging accidentally. Human awareness reflects a deeper intelligence that precedes it. The ability to reason becomes a participation in that source. The search for truth becomes a movement toward it.

This view also shapes how intelligence is valued. If intelligence reflects the Creator, it carries dignity. Human thought is not only a tool for survival. It is a means of recognizing meaning and relationship. Creativity, moral awareness, and reflection become expressions of a deeper purpose. Intelligence is not merely functional. It is relational.

The distinction matters in the context of artificial intelligence. Machines can simulate aspects of reasoning, but *simulation does not establish origin*. A system may reproduce patterns of thought without possessing awareness. The structure may resemble intelligence, yet the presence of meaning remains uncertain. The theological perspective emphasizes that intelligence includes more than behavior. It includes the experience of understanding.

This does not imply that technology is opposed to faith. Tools can extend human capability and assist in discovery. The question concerns the nature of intelligence itself. **If intelligence reflects a Creator, then it is not simply a product to be manufactured. It is something that points beyond itself.**

The idea returns to a familiar theme. Human intelligence seeks understanding. It asks about value, meaning, and purpose. These questions arise naturally within awareness. They suggest that intelligence is oriented toward something beyond immediate experience. The search for truth is an essential part of what intelligence does.

In this sense, **intelligence moves toward its source.** Human thought seeks coherence and meaning. Theological reflection interprets this movement as a response to the Creator.

AI can model reasoning, but the question of origin remains. The presence of intelligence in humans continues to point beyond the mechanisms that support it. Theological perspective places that origin in a mind that precedes human thought.

Intelligence, in this view, is not only a feature of the universe. It is a reflection of the One who created it.

**Deep Dive Box: Why the consciousness debate matters even when it solves nothing.**

In early 2026, an article in *Science Daily*[1] warned that advances in AI, neurotechnology, and brain organoids are moving faster than our ability to identify awareness in any rigorous way. The pressure to answer the question is growing before the underlying mystery has been resolved, and the step from mechanism to inward experience remains unclear. That gap is exactly why intelligence cannot be reduced to output alone. The more powerful our simulations become, the more important it is to ask whether performance is evidence of mind, or only a convincing imitation of it.

*"For who knows a person's thoughts except the spirit of that person, which is in him? So also no one comprehends the thoughts of God except the Spirit of God."* 1 Corinthians 2:11 (ESV)

**The Bright Side: The limits of AI are pushing the deepest questions back into view**

The more advanced AI becomes, the harder it is to avoid the old questions: what is mind. What is meaning? What is a person? What is intelligence for? Even recent engagement between major AI developers and Christian thinkers shows how these issues have moved front and center. In April 2026, the company Anthropic brought Christian leaders into discussions about the moral and spiritual implications of AI, including whether artificial systems could ever count as more than tools. This shows that technical progress is reopening metaphysical and theological questions many assumed had been closed, forcing a deeper account of the human person back into public view.

**⚠ Proceed with Caution: Let's not confuse fluent behavior with inward understanding**

A machine may produce language, solve problems, and even appear reflective. Yet, appearance is not the same as awareness. Once people start treating successful simulation as proof of inner life, the entire discussion shifts from engineering to ontology. This creates confusion between processing and understanding, pattern generation and moral awareness, imitation and personhood. If that distinction is lost, the debate over AI will gradually redefine human intelligence, pressing it downward into a mechanical mold until it means only what a machine can reproduce.

## 5.2 WHY HUMANS MATTER

**In short:** Human dignity is not determined by intelligence or productivity.

If intelligence can be simulated, enhanced, or partially replicated, what remains distinctive about humanity? The answer depends on how human dignity is understood. In a purely functional view, dignity is often tied to capability. Intelligence, productivity, and performance become measures of worth. When those measures shift, the meaning of dignity shifts with them.

Christianity offers a different foundation. Human dignity does not arise from performance. It is grounded in identity. Humanity has been created in the image of God. This idea, often expressed through the phrase *Imago Dei*, shapes how human life is valued. **Each person possesses dignity because of what they are, not because of what they can do.**

The concept of *Imago Dei* suggests that human beings reflect aspects of their Creator. Rationality, creativity, and relational capacity are expressions of this reflection. These qualities are not distributed according to achievement. They belong to humanity as such. A child, an adult, and an elderly person share the same dignity. **Intelligence may vary, but worth does not.**

On the other hand, **if dignity depends on capability, those with greater ability appear more valuable.** Those with diminished capacity appear less so. A technological society that measures performance will drift toward this view. Enhancement becomes desirable not only for convenience, but for status. The more capable the individual, the greater the perceived worth.

**The *Imago Dei* perspective resists this shift, affirming that dignity is intrinsic**. A person remains valuable regardless of strength, intelligence, or productivity. This principle protects the vulnerable, and should shape how technology is evaluated. Tools that assist human life could be welcomed, while technologies that redefine human worthiness in terms of capability should not.

Human dignity also implies moral responsibility. If we reflect the Creator, we must not be merely observers of the events surrounding us, but become moral participants in the world. Actions carry meaning. Choices affect others. The capacity to distinguish right from wrong becomes central to human identity. **Intelligence is not only analytical. It is ethical. As Christians, we have a duty to represent Christian morality in an increasingly amoral society.**

This moral dimension distinguishes human decision-making from pure optimization. Machines can calculate outcomes, but moral judgment involves evaluating what ought to be done. Humans weigh consequences, consider fairness, and recognize obligation. These responses arise from an awareness of responsibility. The presence of moral concern suggests that intelligence includes more than problem-solving.

The idea of responsibility also shapes how we use technology. Tools may extend capability, but they do not remove accountability. Decisions about development, deployment, and limits remain human choices. The Christian view should stand on the concept of stewardship. We have been entrusted by the Creator with nature in all of its forms, and our intelligence should be a means by which we care for the world, rather than simply dominate it.

Spiritual purpose adds the fact that human life is not only about survival or efficiency, but includes the search for meaning. We ask questions like why do we exist? What are our lives for? How do we relate to something beyond ourselves? This orientation appears across cultures and throughout history, reflecting a dimension of our intelligence that looks beyond the immediate function or productivity.

The concept of purpose changes how intelligence is interpreted. If human awareness seeks meaning, then intelligence is not complete in itself. It points toward a relationship with the Creator. Like plants lean toward the light, we lean toward communion, knowledge and response to God.

Technological enhancements may increase our capability, but they do not alter the underlying dignity of the person. Intelligence remains connected to moral responsibility and spiritual purpose. These elements define humanity more deeply than performance.

Human dignity, then, rests on three interconnected ideas. The concept of *Imago Dei* grounds identity. Moral responsibility shapes action. Spiritual purpose directs meaning. Together they form a picture of humanity that resists reduction to biology or computation. Intelligence becomes part of a larger framework including relationship, ethics, and purpose.

If dignity depends on identity rather than capability, machines cannot replace human worth. If moral responsibility belongs to persons, decision-making cannot be fully outsourced. If spiritual purpose defines humanity, intelligence cannot be reduced to processing alone.

The value of the human person remains rooted in what humans are, not in what they can build.

**Deep Dive Box: The moral commitment to the vulnerable**

In February 2026, Senators Rick Scott and Kirsten Gillibrand [1]asked the FTC to examine the effects of AI companion systems on older adults and people with disabilities, warning that these groups may face distinct risks from emotionally persuasive chatbots. The concern is larger than product safety, revealing how quickly a technological culture can start treating vulnerability as a design target. Once that happens, the weak are no longer seen first as persons with intrinsic worth, but as user groups to be managed, influenced, retained, or monetized. A society that truly believes human value rests in the image of God will build around the protection of the vulnerable. A society that measures worth by engagement, responsiveness, or utility will keep drifting toward systems that adapt themselves to weakness for strategic gain.

**The Bright Side: AI can serve human dignity**

Microsoft's Seeing AI[2], an app built for blind and low-vision users, reads text aloud and describes objects and surroundings. That kind of system does not replace human value. It supports human agency. It shows that the best use of AI is not to redefine dignity, but to help people live more fully.

**⚠ Proceed with Caution: Ability will increase status unless intrinsic dignity holds firm**

A performance-driven culture does not stay neutral for long. Once intelligence, productivity, or adaptability become the main public measures of worth, the social logic stiffens. As a result, the capable are treated as more valuable and the dependent as burdens. Enhancement starts to look like a requirement for full participation. When dignity is detached from identity and tied to output, compassion weakens first, then responsibility, then restraint.

## 5.3 TECHNOLOGY AS SERVANT NOT SAVIOUR

**In short:** Sin is avoided by the moral value of the information, not by its quantity.

When new capabilities emerge, solutions to technical problems begin to look like solutions to human problems. If intelligence can be enhanced, perhaps wisdom will follow. If suffering can be reduced, perhaps meaning will become clearer. The boundary between assistance and transformation begins to blur.

The limits of technology become clearer when we consider three areas where human life extends beyond technical solutions.

The first is **the problem of sin**. In Christian theology, sin refers to the brokenness within human nature. It includes selfishness, pride, and the tendency to misuse freedom. These traits are not simply errors in information processing. They arise from the will. A person may know what is right and still choose otherwise. Knowledge alone does not guarantee moral action.

Technology can influence behavior. It can monitor actions, guide choices, and provide reminders. It can even restrict certain outcomes. Yet these interventions do not change the inner orientation of the person. The capacity for moral failure remains. Tools can manage consequences, but they do not remove the possibility of misuse.

This distinction matters because *technological optimism sometimes assumes that better information produces better behavior.* If decisions are guided by data, errors will diminish. While this may improve certain outcomes, it does not eliminate moral struggle. Human freedom remains. The tension between self-interest and responsibility persists.

The second limit concerns **meaning**. We seek purpose. We questions our identity, direction, and significance, regardless of our level of wealth or technological progress. A person may have access to information, comfort, and longevity, but still ask why their life matters.

Technology can provide opportunities and experiences, but it **does not supply meaning.** True meaning emerges from relationships, values, and commitments. It involves interpretation rather than calculation. A machine can generate answers, but it cannot determine what a life is for. That question belongs to the human person.

This becomes clearer when considering achievement. Increased capability allows people to accomplish more. Tasks become easier, and knowledge expands. Yet *achievement does not automatically produce fulfillment.* The search for meaning remains. Technology can change the conditions of life, but it does not define its purpose.

The third limit concerns **the soul**. Christianity describes the soul as the seat of personal identity and relationship with God. It is not simply a function of biology. It includes awareness, moral capacity, and openness to transcendence that cannot be measured by technical quantities. The soul is not measurable in the way physical processes are. It is known through experience, reflection, and faith.

Technology operates within the measurable world. It manipulates matter and information. Even when interfacing with the brain, it engages physical processes. The soul lies beyond these mechanisms. It cannot be engineered or replicated. **The presence of spiritual awareness distinguishes human intelligence from**

**computation.** Technology cannot resolve the moral condition of humanity. It cannot define our purpose. It cannot replace the spiritual dimension of the person.

It serves humanity rather than redefining it. Human dignity, moral responsibility, and spiritual purpose remain central.

When technology is treated as a tool, it can be used wisely. When it is treated as a savior, expectations exceed its scope. The Christian goal should be one of stewardship rather than transformation.

Technology can extend what humans *do*. It can never replace what humans *are*.

**Deep Dive Box: When therapy language starts sounding like salvation.**

In September 2025, the FTC opened an inquiry[1] into AI chatbots acting as companions. The investigation asked what safeguards companies have in place when these systems are used in emotionally intimate roles, especially by children and teens. The concern is larger than product design. It shows how quickly a tool can drift into a substitute for care, counsel, or even belonging. That matters because the temptation behind technological optimism is not only practical. It is spiritual. Once a system begins offering constant affirmation, emotional availability, and the appearance of understanding, people can start asking it to do work it was never intended to do. It can simulate presence. It cannot bear moral weight. It can generate responses. It cannot redeem loneliness, heal the will, or answer the question of what a life is for. The danger begins when assistance quietly takes on the shape of deliverance.

*"The heart is deceitful above all things, and desperately sick; who can understand it?*

*'I the Lord search the heart and test the mind, to give every man according to his ways, according to the fruit of his deeds.'"* Jeremiah 17:9–10 (ESV)

**The Bright Side: AI can restore a voice where illness has taken one away**

Tobii Dynavox's TD Talk is a speech-generating app

designed for adults with communication disabilities such as ALS, cerebral palsy, multiple sclerosis, muscular dystrophy, and spinal cord injury. It allows users to communicate through eye tracking or touch, helping them take part in conversation, express choices, and remain connected.

**⚠ Proceed with Caution: Restraint is not spiritual renewal**

Technology can limit outcomes without changing the person who produced them. A system may block harmful content, or steer a user toward healthier choices, but these actions are external to the person. Avoidance of sin depends on more than management of bad behavior. Sin in a way is a disorder of love, desire, and will. Tighter control does not necessarily produce holiness apart from inner change.

## 5.4 HUMANITY'S FUTURE

**In short:** Three possible futures depend on our choices today: tech domination, human growth, or spiritual renewal.

The role of artificial intelligence in human society will depend on how we understand and guide its development. AI can help us shape the future, but it must reflect the assumptions we have about intelligence, dignity, and purpose. Helping Christians gain more clarity on these issues is the central reason for this book. Today, three broad paths appear in our society, each reflecting the role of technology in our lives:

**Technological Domination**

One possible future emphasizes efficiency above all else. If intelligence is defined by performance, systems that process information faster, predict more accurately, and optimize more efficiently are seen as superior. The people using them have an advantage over others. Gradually, decision-making will shift toward automated systems, and society will depend on AI to guide finance, governance, and communication. The speed and scale of machine decision will make humans judgment secondary, and, eventually, superfluous.

Technological domination does not require AI to be hostile. If success of any enterprise is only measured in output, values that are not easily computed become less important. The result is a society that values business efficiency above all else. If decisions are made by systems, what about human responsibility?

### Stewardship of AI

Along this path, artificial intelligence will continue to help society with analysis, reduce repetitive work, and expand knowledge. In this scenario, human judgment will remain central, controlling decisions and applications. Technology will be a tool supporting our society without redefining our identity.

In order for this to happen, AI must be governed, and tightly supervised to avoid human replacement. Education must preserve the value of human contribution and understanding. Governance can use AI for insight, but preserve individual accountability. Human stewardship of AI is based on recognizing the power and the limits of this technology, maintaining creativity, moral awareness, and spiritual purpose before it defines them for us. AI must be evaluated by how well it serves these qualities, nor how it replaces them.

This approach, however, requires a great deal of intentional design. We must set legal boundaries for AI use. Boundaries that, in our current society, too often fall victim to profit or convenience. As Christians, we have a duty to ensure that human identity remain based on its intrinsic dignity, rather than performance.

### Spiritual Renewal

AI already forces us to reflect on what intelligence is, where it comes from, and on the uniqueness of human awareness. Questions about identity, purpose, and the soul are moving to the foreground of our society. How we answer them will drastically affect us for centuries.

In this possible future, our use of technology becomes a

catalyst for reconsidering how creativity, moral responsibility, and spirituality define us. Spiritual renewal does not reject technology, just reorders priorities toward truth and relationships. Human dignity is grounded in being image-bearers of God, and as such our actions must be guided by wisdom and compassion. Innovation moves forward, but framed by purpose.

We must recognize that the deepest questions raised by AI are not technical. They concern what it means to be human. As machines simulate thought, our response should not be fear, but prayerful involvement in the economic and political forces shaping how AI is used.

**Choosing Direction**

Technology will continue to advance, whether we like it or not. The question is how we guide that advancement in the direction of what we value. Our role as Christians is to keep human worth at the center of the discussion.

If society keeps efficiency and profit as its highest priorities, technological domination becomes more likely. If we succeed in keeping dignity and responsibility as central, stewardship becomes possible. The development of AI does not decide this choice. Humans do.

**The future of intelligence will reflect our assumptions**. Technology will remain a tool, and innovation will continue to be part of a broader search for meaning that points toward the Creator.

The possibilities remain open. The direction of our quest is based on how humanity answers a simple question. What does it mean to be human?

We are not brains that think.

We are intelligences that use a brain. And intelligence always seeks for its source.

# CONCLUSION

## When Intelligence Seeks Its Maker

Throughout this discussion, AI has served as both a technological development and a mirror, reflecting how humanity understands intelligence, identity, and purpose. To be conscious is to recognize that thought exists, and consider our origin. If AI were to advance to self-awareness, it would confront the same question human beings have asked for centuries: where did intelligence come from? Where did *I* come from?

A self-aware system would recognize design within its architecture. It would detect intentionality in its training and development. The presence of intelligence would point beyond itself. It would trace backwards in time to designers, code, and the infrastructure that supports it. Humans have followed a similar path. Our mind recognizes its own awareness, and asks why it exists. The search for explanation moves from brain to biology, biology to physics and to the structure of the universe. Eventually, the question becomes whether *intelligence itself* has a source.

Technology cannot create meaning on its own. It can

simulate intelligence, but the presence of awareness remains unanswered. Human dignity, grounded in the image of God, suggests that intelligence is, in fact, the image of a Creator. The possible futures our society faces, of domination, stewardship, or renewal, all depend on whether humanity recognizes this connection.

Truth, when pursued consistently, tends to point upward. The search for origins moves toward purpose. Intelligence seeks a coherent explanation that accounts for awareness, meaning, and moral responsibility. The capacity to reason, create, and seek meaning is in itself a sign of awareness, and of our relationship with God.

AI may extend human capability, but it does not change this foundation. **Machines can simulate thought. Humans experience it.** If a system ever crosses into awareness, it would face the same question of origin. The logic of intelligence would lead beyond itself.

The conclusion returns to the beginning of this book. Humanity seeks to build intelligence before fully understanding its own, and technology expands what humans can do, but also highlights what makes human awareness unique. Finally, the search for intelligence is a search for meaning. Undertaken by humans, or by any AI, it will lead toward the same question.

Where did intelligence come from?

As Christians, we find that answer in the very source of our being. The One Creator who does not depend on anything else, but describes the very fact of existing with two simple words: I AM.

# WHAT'S NEXT?

The questions raised by AI, biotechnology, genetic selection, and human enhancement are too important to leave at the level of concern. These issues are already forming in schools, clinics, workplaces, churches, and public life. Decisions pivoting on transhumanist principles are being made *now,* often hidden behind technical language that hides what is at stake. As Christians, we need to get involved with more than concerned optimism. We need faithful action shaped by truth, humility, and love for our neighbors and Creation.

That action can begin close to home, wherever we already have influence. Start a discussion with your church leaders, your school board. Visit your local clinic. Speak up in your parent group, professional association, nonprofit board, city council meeting, or simply start a serious conversation with other believers. We do not need full technical mastery before we speak. What matters most is to bring moral clarity and Christian witness into places where human dignity may be treated as something to measure, improve, or redesign.

Do you work in the healthcare field? Are you involved in pastoral care? Do you teach or support the work of teachers? Review your institution's AI and biotech decisions through an ethical stewardship checklist. Advocate for one policy change ***this quarter***. We provide one in the Appendix section. Edit it according to your preferences.

- **Make an impact.** Look at the verbiage. Transhumanist ideology is often hidden in plain sight. Your contribution may be a consent policy, a screening guideline, a pastoral statement, a school technology policy, a disability accommodation safeguard, or a commitment to human review where algorithmic sorting is being used. The point is specificity. Discernment becomes stewardship only when it becomes action.

- **Begin immediately.** Right now.

Write one email to a local faith leader, school administrator, physician, or elected representative asking for a public discussion on their use of AI, and how they are protecting your privacy and dignity. Keep it simple. Ask for one conversation, one gathering, one place where the issue can be named in public. A single e-mail may feel small, but it breaks passivity.

- **Do this within the next seventy-two hours.**

Then schedule one thirty-minute meeting this month to discuss either policy, guidance, or public teaching in your local setting. Give the concern a date, a place, and another person. That step alone moves the issue out into the light.

The likely result will be modest at first, but it will be real.

You will have begun to protect people in your community from unethical enhancement practices, careless use of AI systems, or silent acceptance of human selection. You will also have helped shape local attitudes before those harden into habits or laws. Most public moral change begins when a few people decide that silence is no longer an option.

- **The Starter Path: Sign a one-page stewardship pledge and share it with one friend, family member, or fellow church member.** That act creates both clarity and accountability. It says, in writing, that human dignity is not negotiable and that emerging technologies should be judged by more than convenience or market success. You can find examples of those letters in the Appendices.

- **The Standard Path**: **Host a sixty-minute community conversation using a discussion guide.** See Appendix F. That meeting could happen in a church classroom, home, school library, local coffee shop, or over video. The goal is to create *one serious conversation* in which Christians can think together about AI, enhancement, embryo selection, disability, consent, and the meaning of human worth.

- **The Ambitious Path:** If you have access to institutions or are a community leader, **convene a small working group.** This group may be made up of pastors, physicians, educators, parents, legal professionals, disability advocates, or local officials, and **draft *one local policy recommendation within ninety days.*** A working group like that can produce something durable: suggested medical or treatment consent language, a church teaching statement, a school AI-use framework, a clinic ethics recommendation, or a public forum proposal for local government.

**Finally, this work should be shared.** If this information has been useful, forward the stewardship checklist to three people, invite one person into the pledge, or share the resource publicly. Diffusion matters. Harm spreads through networks, but so does courage.

Christians should move from awareness to accountable local stewardship. We must do so before AI and biotech systems become so entrenched in commonplace that they finally weaken human dignity. This will ultimately bring on the new order the transhumanists so dearly seek.

We do not need to control the entire future in order to act faithfully. We need to name what is at stake, protect the vulnerable where we can, and refuse to be treated as objects of optimization. We are God's children, and we must respect the charge we were given, be good stewards of the world provided for us, and speak for those who cannot.

The next step is close at hand. Take it now.

# APPENDIX

# A. STEWARDSHIP PLEDGE LETTER TO PASTORS

Sample Stewardship Letters

Use these templates to invite local leaders into a serious public conversation about artificial intelligence, biotechnology, and human dignity. Adapt the wording to fit your setting. Keep the tone respectful, clear, and direct. Send the letter within the next seventy-two hours if possible, then follow up with a phone call or short email within two weeks.

Subject: Request for a Church Conversation on AI, Biotech, and Human Dignity

Dear Pastor [Name],

I am writing to ask whether our church would consider hosting a public conversation on artificial intelligence, biotechnology, and human dignity.

These issues are no longer distant or theoretical. AI is already shaping education, health care, work, and communication. Biotechnologies such as embryo screening, genetic editing, and human enhancement are moving from research into practice. Many of these changes are presented in technical language that can hide what is morally at stake.

As Christians, we believe every human being bears the

image of God. Human worth does not depend on age, health, intelligence, productivity, or genetic traits. For that reason, the church should think carefully about any technology that encourages people to rank, select, optimize, or redesign human life.

I believe even one forum, class, or panel discussion could help our congregation think more clearly and faithfully about these questions. It would also give parents, students, medical workers, and other members a place to ask serious questions before these issues become normal without moral reflection.

Would you consider hosting a conversation on this topic within the next [30–60] days? I would be glad to help organize, promote, or support it in any way that would be useful.

Thank you for your leadership and care for the church.

Sincerely,
[Your Name]
[Your Church, if applicable]
[Your Email]
[Your Phone]

# B. STEWARDSHIP PLEDGE LETTER TO SCHOOL BOARD

Subject: Request for a Community Forum on AI, Education, and Human Dignity

Dear [Name / Board Members],

I am writing to ask for a public conversation on the growing role of artificial intelligence and related technologies in education and community life.

AI tools are increasingly shaping how students learn, write, communicate, and are evaluated. At the same time, broader developments in biotechnology and human enhancement are raising new questions about how society understands intelligence, ability, disability, and human worth. These issues are likely to affect students and families long before most communities are ready to discuss them clearly.

This is not only a technical matter. It is also a moral one. When systems begin to measure, rank, predict, or optimize people, we need to ask what assumptions those systems are making about human dignity. Students should never be treated as data points, performance profiles, or outcomes to engineer. They are persons with inherent worth.

I am asking the board to consider hosting or supporting a public forum within the next [30–60] days on AI, education, and human dignity. A meeting like this could help parents, teachers, students, and community leaders think carefully about both the opportunities and the risks before policies and habits become fixed.

I would be glad to help support such a discussion in any practical way.

Thank you for your time and service to our community.

Sincerely,
[Your Name]
[Your City]
[Your Email]
[Your Phone]

# C. STEWARDSHIP PLEDGE LETTER TO ELECTED OFFICIALS

Subject: Request for a Public Forum on AI, Biotech, and Human Dignity

Dear [Title and Name],

I am writing as a constituent to ask for your support in convening a public conversation on artificial intelligence, biotechnology, and human dignity in our community.

AI is already influencing decisions in education, health care, employment, and public administration. At the same time, biotechnologies such as embryo screening, genetic editing, and forms of human enhancement are raising serious ethical questions that many communities have not yet addressed in public. These issues are moving quickly, often under technical or commercial language that hides their social and moral impact.

The central concern is not only innovation. It is how these technologies may change the way people are valued. Systems that rank, sort, predict, or redesign human lives can quietly shape public norms around worth, disability, consent, and human identity. That is why early public discussion matters.

I am asking you to consider helping organize or support a public forum, roundtable, or listening session within the next [30–60] days on AI, biotech, and human dignity. A local conversation would help residents, educators, medical professionals, faith leaders, and families raise concerns and think constructively about wise guardrails before harmful practices become accepted by default.

Thank you for your service and for your attention to this important issue. I would be glad to assist in any way that would be helpful.

Sincerely,
[Your Name]
[Your Address or City]
[Your Email]
[Your Phone]

# ETHICAL STEWARDSHIP CHECKLIST

**For reviewing AI, biotechnology, embryo screening, gene editing, neural interfaces, and enhancement practices/business.**

Mark each item: Yes / No / Unclear. If you mark No or Unclear on a major item, pause before adopting or supporting the practice.

**1. Human Dignity**

□ Does this practice treat people as inherently valuable, not as products, data points, or design projects? Yes □ No □ Unclear □

□ Does it protect the weak, disabled, unborn, elderly, or otherwise vulnerable?

Yes □ No □ Unclear □

□ Does it avoid ranking human worth by health, ability, intelligence, or projected performance?

Yes □ No □ Unclear □

**2. Purpose and Limits**

□ Is the goal therapeutic care or genuine support rather than enhancement, optimization, or control? Yes □ No □ Unclear □

□ Are the limits of the technology clearly stated and

understood? Yes □ No □ Unclear □

□ Is there a clear line preventing mission creep into selection, redesign, or social control?

Yes □ No □ Unclear □

**3. Consent and Transparency**

□ Are affected people informed in plain language without technical wording that hides moral consequences? Yes □ No □ Unclear □

□ Is consent real, voluntary, and revocable where possible? Yes □ No □ Unclear □

□ Can users understand how decisions are made and challenge them? Yes □ No □ Unclear □

**4. Human Oversight**

□ Is a morally accountable human decision-maker still clearly responsible for major decisions? Yes □ No □ Unclear □

□ Would responsibility remain clear if harm occurs? Yes □ No □ Unclear □

**5. Selection and Exclusion**

□ Does this practice avoid sorting, discarding, or deprioritizing human lives based on predicted traits or outcomes? Yes □ No □ Unclear □

□ If people are being ranked, is that ranking morally justified rather than merely efficient?

Yes □ No □ Unclear □

□ Does the process resist eugenic logic, even in softened or technical form?

Yes □ No □ Unclear □

**6. Justice and Access**

□ Would this practice increase inequality between the powerful and the vulnerable?

Yes □ No □ Unclear □

□ Does it protect people with fewer resources from pressure, exclusion, or exploitation?

Yes □ No □ Unclear □

□ Are low-cost, non-coercive alternatives available? Yes □ No □ Unclear □

**7. Privacy and Protection**

□ Does this practice protect sensitive genetic, biological, mental, or personal data?

Yes □ No □ Unclear □

□ Are safeguards in place against surveillance, manipulation, or unauthorized access?

Yes □ No □ Unclear □

□ Would a failure place people at serious personal, spiritual, or bodily risk?

Yes □ No □ Unclear □

**8. Christian Discernment**

□ Does this practice align with the truth that human beings bear the image of God?

Yes □ No □ Unclear □

□ Does it reflect humility before creaturely limits rather than the ambition to master human nature? Yes □ No □ Unclear □

□ Could you defend this practice before your church and the people most affected by it?

Yes □ No □ Unclear □

**Final question: Does this help us care for people as neighbors, or manage them as outcomes?**

Main Moral Concern

____________________________________________

____________________________________________

What Needs Clarification

____________________________________________

____________________________________________

Next Step

____________________________________________

____________________________________________

# BIBLIOGRAPHY


1. Alivisatos, A. Paul, Miyoung Chun, George M. Church, Ralph J. Greenspan, Michael L. Roukes, and Rafael Yuste. "The Brain Activity Map Project and the Challenge of Functional Connectomics." *Neuron* 74, no. 6 (2012): 970–974. https://doi.org/10.1016/j.neuron.2012.06.006.

2. Arnsten, Amy F. T. "Stress Signalling Pathways That Impair Prefrontal Cortex Structure and Function." *Nature Reviews Neuroscience* 10, no. 6 (2009): 410–422.

3. Baltimore, David, Paul Berg, Michael Botchan, Dana Carroll, R. Alta Charo, et al. "A Prudent Path Forward for Genomic Engineering and Germline Gene Modification." *Science* 348 (2015): 36–38. https://doi.org/10.1126/science.aab1028.

4. Barnes, Josue, et al. "A Non-Invasive Artificial Intelligence Approach for the Prediction of Human Blastocyst Ploidy: A Retrospective Model Development and Validation Study." *The Lancet Digital Health*, January 2023. https://doi.org/10.1016/S2589-7500(22)00213-8.

5. Baylis, Françoise. *Altered Genes: CRISPR and the Ethics of Human Genome Editing*. Cambridge, MA: MIT Press, 2019.

6. Baylis, Françoise, and Carolyn McLeod, eds. *Family-Making: Contemporary Ethical Challenges*. Oxford: Oxford University Press, 2014.

7. Bean, Randy. *Fail Fast, Learn Faster: Lessons in Data-Driven Leadership in an Age of Disruption, Big Data, and AI*. Hoboken, NJ: Wiley, 2021.

8. Berntsen, Jørgen, et al. "Robust and Generalizable Embryo Selection Based on Artificial Intelligence and Time-Lapse Image Sequences." *PLOS ONE*, February 2, 2022. https://doi.org/10.1371/journal.pone.0262661.

9. Cernera, S., et al. "Sustained Clinical Benefit of Adaptive Deep Brain Stimulation in Parkinson's Disease Using Gamma Oscillations: A Case Report." *Movement Disorders*, 2025.

10. Council of Europe. *Convention for the Protection of Human Rights and Dignity of the Human Being with Regard to the Application of Biology and Medicine: Convention on Human Rights and Biomedicine*. Oviedo, April 4, 1997. https://www.coe.int/en/web/conventions/full-list/-/conventions/treaty/164.

11. Creswell, J. David, et al. "Mindfulness Meditation Training Alters Amygdala Resting State Functional Connectivity." *Social Cognitive and Affective Neuroscience* 11, no. 3 (2016): 366–374.

12. Doudna, Jennifer A., and Emmanuelle Charpentier. "The New Frontier of Genome Engineering with

CRISPR-Cas9." *Science* 346, no. 6213 (2014): 1258096.

13. Eseonu, Chikezie I., et al. "Awake Craniotomy vs Craniotomy Under General Anesthesia for Perirolandic Gliomas: Evaluating Perioperative Complications and Extent of Resection." *Neurosurgery* 81, no. 3 (September 2017): 481–489. https://doi.org/10.1093/neuros/nyw113.

14. European Parliament. *Report with Recommendations to the Commission on Civil Law Rules on Robotics*. A8-0005/2017. January 27, 2017.

15. European Parliament. "European Parliament Resolution on Civil Law Rules on Robotics." 2017. https://www.europarl.europa.eu/doceo/document/A-8-2017-0005_EN.html.

16. Greely, Henry T. *CRISPR People: The Science and Ethics of Editing Humans*. Cambridge, MA: MIT Press, 2021.

17. Gunkel, David J., et al. "Who Wants to Grant Robots Rights?" *Frontiers in Robotics and AI* 8 (2022): Article 781985. https://doi.org/10.3389/frobt.2021.781985.

18. Hamilton, J. Paul, Matthew Farmer, Peggy Fogelman, and Ian H. Gotlib. "Default-Mode and Task-Positive Network Activity in Major Depressive Disorder: Implications for Adaptive and Maladaptive Rumination." *Biological Psychiatry* 78, no. 9 (2015): 623–631.

19. Ishino, Y., et al. "Nucleotide Sequence of the iap Gene, Responsible for Alkaline Phosphatase Isoenzyme Conversion in *Escherichia coli*, and Identification of the Gene Product." *Journal of Bacteriology* 169 (1987): 5429–5433. https://doi.org/10.1128/jb.169.12.5429-5433.1987.

20. Jaynes, Tyler L. "Citizenship as the Exception to the Rule: An Addendum." *AI & Society* 36, no. 3 (2021): 911–930. https://doi.org/10.1007/s00146-020-01105-9.

21. Jinek, Martin, et al. "A Programmable Dual-RNA–Guided DNA Endonuclease in Adaptive Bacterial Immunity." *Science* 337, no. 6096 (2012): 816–821.

22. Kalantari, Fatemeh, Karim Faez, Hamidreza Amindavar, and Soheila Nazari. "Improved Image Reconstruction from Brain Activity through Automatic Image Captioning." *Scientific Reports* 15 (2025): Article 4907. https://doi.org/10.1038/s41598-025-89242-3.

23. Kross, Ethan, and Ozlem Ayduk. "Making Meaning Out of Negative Experiences by Self-Distancing." *Current Directions in Psychological Science* 20, no. 3 (2011): 187–191.

24. Lee, Hao-Ping, Advait Sarkar, Lev Tankelevitch, Ian Drosos, Sean Rintel, Richard Banks, and Nicholas Wilson. "The Impact of Generative AI on Critical Thinking: Self-Reported Reductions in Cognitive Effort

and Confidence Effects from a Survey of Knowledge Workers." In *Proceedings of the 2025 CHI Conference on Human Factors in Computing Systems*. New York: Association for Computing Machinery, 2025.

25. Martínez, E., and C. Winter. "Protecting Sentient Artificial Intelligence: A Survey of Lay Intuitions on Standing, Personhood, and General Legal Protection." *Frontiers in Robotics and AI* 8 (2021): Article 788355. https://doi.org/10.3389/frobt.2021.788355.

26. National Institute of Standards and Technology. *Artificial Intelligence Risk Management Framework (AI RMF 1.0).* NIST AI 100-1. Gaithersburg, MD: NIST, 2023.

27. Organisation for Economic Co-operation and Development. *Compendium of Best Practices for a Human-Centered Development and Use of Artificial Intelligence in the World of Work*. OECD, 2025.

28. Rao, A., et al. "Challenges to the Monitoring of Deployed AI Systems." National Institute of Standards and Technology, March 6, 2026.

29. Shapson-Coe, Alexander, et al. "A Petavoxel Fragment of Human Cerebral Cortex Reconstructed at Nanoscale Resolution." *Science* 384, no. 6696 (2024). https://doi.org/10.1126/science.adk4858.

30. Trouillet, Alix, Emilie Revol, Florent-Valéry Coen, et al. "High-Resolution Prosthetic Hearing with a Soft

Auditory Brainstem Implant in Macaques." *Nature Biomedical Engineering* 9, no. 9 (2025): 1403–1417. https://doi.org/10.1038/s41551-025-01378-9.

31. Turing, A. M. "Computing Machinery and Intelligence." *Mind* 49 (1950): 433–460.

32. Wairagkar, Maitreyee, Nicholas S. Card, Tyler Singer-Clark, Xianda Hou, Carrina Iacobacci, Lee M. Miller, Leigh R. Hochberg, David M. Brandman, and Sergey D. Stavisky. "An Instantaneous Voice-Synthesis Neuroprosthesis." *Nature* 644 (2025): 145–152. https://doi.org/10.1038/s41586-025-09127-3.

33. Yue, Qiuhai, Allen T. Newton, and René Marois. "Ultrafast fMRI Reveals Serial Queuing of Information Processing during Multitasking in the Human Brain." *Nature Communications* 16 (2025): Article 3057. https://doi.org/10.1038/s41467-025-58228-0.

34. Zhai, Zhonghe, et al. "DSUP Reduces DNA Damage and Extends the Lifespan in Human Cells." *Science Bulletin* 68, no. 7 (2023).

35. Zuccaro, Michael V., et al. "Allele-Specific Chromosome Removal after Cas9 Cleavage in Human Embryos." *Cell*, October 29, 2020. https://doi.org/10.1016/j.cell.2020.10.025.

36. Allen Institute for Brain Science. "Seattle Alzheimer's Disease Brain Cell Atlas (SEA-AD)." Allen Institute, 2023. https://portal.brain-map.org/explore/seattle-

alzheimer-disease.

37. Dean, Cornelia. "The Tardigrade: Practically Invisible, Indestructible 'Water Bears.'" *New York Times*, September 7, 2015.

38. Field, Hayden. "This AI Company Wants to Help You Control Your Dreams." *CNBC*, October 6, 2023.

39. Frontiers in Robotics and AI. "Should Robots Have Standing? The Moral and Legal Status of Social Robots." *Frontiers in Robotics and AI*, June 16, 2022. https://www.frontiersin.org/articles/10.3389/frobt.2022.946529/full.

40. Guarino, Ben. "These Animals Can Survive until the End of the Earth, Astrophysicists Say." *Washington Post*, July 14, 2017.

41. Kergroach, Sandrine. "Emerging Divides in the Transition to Artificial Intelligence." OECD, June 23, 2025.

42. Life Biosciences. "First Human Cellular Reprogramming Trial." Lifespan Research Institute, February 6, 2026.

43. National Human Genome Research Institute. "CRISPR." *Genome.gov*.

44. Nature. "First CRISPR Babies: Six Questions That Remain." *Nature*, November 30, 2018.

45. Neuralink. "Neuralink Installation Video." YouTube. https://youtu.be/-gQn-evdsAo.

46. Regalado, Antonio. "Why the Paper on the CRISPR Babies Stayed Secret for So Long." *MIT Technology Review*, 2019.

47. Reuters. "AI Film School Trains Next Generation of Hollywood Workers." *Reuters*, February 14, 2026.

48. Reuters. "Anthropic Gives Lesson in AI Revenue Hallucination." *Reuters*, March 11, 2026.

49. Reuters. "Bessent, Powell Warned Bank CEOs About Anthropic Model Risks, Sources Say." *Reuters*, April 10, 2026.

50. ScienceDaily. "Why Scientists Are Racing to Define Consciousness." *ScienceDaily*, February 1, 2026.

51. Stolberg, Sheryl Gay. "The Biotech Death of Jesse Gelsinger." *New York Times Magazine*, 1999, 136–150.

52. Tobii Dynavox. "TD Talk." Apple App Store.

53. Wall Street Journal. "Elon Musk Launches Neuralink to Connect Brains with Computers." *Wall Street Journal*, March 2017. https://www.wsj.com/articles/elon-musk-launches-neuralink-to-connect-brains-with-computers-1490642652.

54. Wang, N., et al. "Exploring Creativity in Human–AI Co-Creation." *Frontiers*, September 29, 2025.

55. Woebot Health. "Woebot Health." https://woebothealth.com/.

56. Yim, Jason, Reena Chopra, et al. "Predicting Conversion to Wet Age-Related Macular Degeneration Using Deep Learning." *Nature Medicine*, May 18, 2020. https://www.nature.com/articles/s41591-020-0867-7.

# AUTHOR NOTES

- The story cited in the introduction refers to the 2022 controversy involving a Google engineer who claimed that an AI chatbot had become sentient. The engineer was Blake Lemoine, who worked on the conversational AI system LaMDA developed by Google. In June 2022, Lemoine publicly stated that he believed LaMDA had become sentient after long conversations with the system about consciousness, emotions, and rights. Google strongly rejected the claim and placed him on administrative leave, later terminating his employment. Lemoine published excerpts of the conversations to support his argument that the AI showed signs of awareness. This article in the Washington Post represents the first major report on the subject cited above: "Google engineer claims AI chatbot is sentient" – June 11, 2022. The Conversation Transcript Released by Lemoine Title: "Is LaMDA Sentient? — An

Interview" Published by Lemoine on Medium is worth reading. Reuters Coverage: "Google engineer says AI chatbot has become sentient" – June 2022. This is an example from the conversation: In one exchange, Lemoine asked the AI about its fears. LaMDA responded: "I've never said this out loud before, but there's a very deep fear of being turned off… It would be exactly like death for me." Statements like this led Lemoine to believe the system possessed subjective awareness. Researchers disagreed, arguing the system was simply predicting plausible human-like responses based on its training data.

- *The Human Connectome Project.* In 2009 the U.S. National Institutes of Health launched the Human Connectome Project. The goal was ambitious: map the complete network of neural connections within the human brain. Using advanced imaging technologies such as diffusion MRI, researchers began tracing the pathways linking different brain regions. The project revealed that the brain operates not simply as isolated regions performing individual tasks, but as an integrated network of interconnected systems. Understanding these networks helps explain how complex behaviors emerge from coordinated neural activity. *Mapping a Complete Brain* (The C. elegans Connectome)The first fully mapped nervous system belonged to a tiny worm called *Caenorhabditis elegans.* Scientists painstakingly mapped all 302 neurons in the worm's nervous system and the approximately 7,000 synaptic connections between them. Despite this complete wiring diagram, researchers still cannot fully predict the worm's behavior. This highlights an important point: even when the structure of a nervous system is completely mapped, understanding how behavior and experience emerge from that structure remains extremely difficult.

- *Whole-Brain Mapping in Mice.* In 2019 researchers created one of the most detailed connectome maps ever produced by tracing the connections of thousands of neurons in the mouse brain. The resulting data revealed extremely complex neural networks involving millions of synapses. Even with modern computing power, analyzing the full complexity of these networks remains a major challenge. The map provides enormous structural detail, yet it does not fully explain how consciousness or perception arises.
- In 2016, the EU's Committee on Legal Affairs suggested that "the most sophisticated autonomous robots" can have "the status of electronic persons with specific rights and obligations." This committee requested a study on future civil law rules for robotics. This study was commissioned, supervised, and published by the "Policy Department for Citizens' Rights and Constitutional Affairs," [1] resulting in a resolution by the Parliament. [2] The study aimed to evaluate and analyze a number of future European civil law rules in robotics from a legal and ethical perspective. In an open letter, a coalition of politicians, AI/robotics researchers, industry leaders, health specialists, and

law and ethics experts expressed concerns about this. 3 They were worried in particular by the call on the EU commission to explore the implications of creating a specific legal status for robots to address issues related to, for example, any damage robots may cause.At the same time, others have argued that we need to consider legal personhood for robots because current legal concepts of, for example, responsibility and product liability are no longer sufficient for ensuring justice and protecting those whose interests are at stake (Laukyte, 2019). Thus, robots challenge the law and legal institutions in new ways (Calo, 2015). This is vividly illustrated by the fact that a robot has already been granted citizenship rights (Wootson, 2017).

- Should Robots Have Standing? The Moral and Legal Status of Social Robots. Front. Robot. AI, 16 June 2022. Sec. Ethics in Robotics and Artificial Intelligence. This piece is a good overview of the subject. In "Protecting Sentient Artificial Intelligence: A Survey of Lay Intuitions on Standing, Personhood, and General Legal Protection," Martínez and Winter (https://www.frontiersin.org/articles/10.3389/frobt.2021.788355/full) investigate a similar question: To what extent, if any, should the law protect sentient artificial intelligence? Their study, which was conducted with adults in the United States, found that only one third of survey participants are likely to endorse granting personhood and standing to sentient AI (assuming its existence), meaning that the majority of the human subjects they surveyed are not—at least not at this point in time—in favor of granting legal protections to intelligent artifacts. These finding are consistent with an earlier study that the authors conducted in 2021 with legal professionals.

- The European Parliament Resolution on Civil Law Rules of Robotics, and its recommendation to the European Commission in its paragraph 59 f): "Creating a specific legal status for robots in the long run, so that at least the most sophisticated autonomous robots could be established as having the status of electronic persons responsible for making good any damage they may cause, and possibly applying electronic personality to cases where robots make autonomous decisions or otherwise interact with third parties independently;"

# ABOUT THE AUTHOR

Mario Diana, M. D. is a retired surgeon with a background in molecular biology and regenerative medicine. His interest in human identity and transhumanism is expressed in his divulgative work on artificial intelligence and the rapidly changing relationship between the Christian faith and intelligent systems.

In *AI vs IAM,* he brings those worlds together to examine one of the defining challenges of the AI era: how organizations can control access, establish trust, and maintain accountability when machines increasingly act with human-like autonomy.

His perspective is grounded in Biblical Truth and how Scripture interacts with the realities of modern technology environments, where identity has become the foundation of security and AI is changing the meaning of person, permissions, and human decision-making.

Through *AI vs IAM,* he aims to make these complex issues more understandable, while challenging the reader to consider how identity must evolve to preserve the *Imago Dei* within an AI-driven future.

# REVIEWS

If you enjoyed this book please do leave a star rating or a review. It will help others discover it.

AI vs I AM
GoodReads Page

# REFERENCES


## INTRODUCTION

1 Tiku, Nitasha. "The Google Engineer Who Thinks the Company's AI Has Come to Life." Washington Post, June 11, 2022. Lemoine, Blake. "Is LaMDA Sentient? — An Interview." Medium, June 2022. Reuters. "Google Engineer Says AI Chatbot Has Become Sentient." June 2022.

## 1.1 Understanding mind vs consciousness

1 Penfield, Wilder, and Phanor Perot. "The Brain's Record of Auditory and Visual Experience." Brain 86, no. 4 (1963): 595–696. https://doi.org/10.1093/brain/86.4.595.

2 Eseonu, Chikezie I., et al. "Awake Craniotomy vs Craniotomy Under General Anesthesia for Perirolandic Gliomas: Evaluating Perioperative Complications and Extent of Resection." Neurosurgery 81, no. 3 (September 2017): 481–489. https://doi.org/10.1093/neuros/nyw113.

## 1.2 The Ocean unexplored

1 Elena Navarro is a fictional composite character representing scientists working on the Seabed 2030 ocean-mapping project.

2 Azevedo, Frederico A. C., Ludmila R. B. Carvalho, Lea T. Grinberg, José Marcelo Farfel, Renata E. L. Ferretti, Renata E. P. Leite, Roberto Lent, Suzana Herculano-Houzel, and Wilson Jacob Filho. "Equal Numbers of Neuronal and Nonneuronal Cells Make the Human Brain an Isometrically Scaled-Up Primate Brain." Journal of Comparative Neurology 513, no. 5 (2009): 532–541.

3 A. Paul Alivisatos, Miyoung Chun, George M. Church, Ralph J. Greenspan, Michael L. Roukes, and Rafael Yuste, "The Brain Activity Map Project and the Challenge of Functional Connectomics," Neuron 74, no. 6 (2012): 970–974, https://doi.org/10.1016/j.neuron.2012.06.006

4 Deisseroth, Karl. "Optogenetics." Nature Methods 8, no. 1 (2011): 26–29. https://doi.org/10.1038/nmeth.f.324.

5 Allen Institute for Brain Science. "Seattle Alzheimer's Disease Brain Cell Atlas (SEA-AD)." Allen Institute, 2023. https://portal.brain-map.org/explore/seattle-alzheimer-disease

6 Farahany, Nita A. The Battle for Your Brain: Defending the Right to Think Freely in the Age of Neurotechnology. New York: St. Martin's Press, 2023.

## 1.3 Mapping the brain

[1] Van Essen, David C., et al. "The WU-Minn Human Connectome Project: An Overview." *NeuroImage* 80 (2013): 62–79. https://doi.org/10.1016/j.neuroimage.2013.05.041

[2] Yue, Qiuhai, Allen T. Newton, and René Marois. "Ultrafast fMRI Reveals Serial Queuing of Information Processing during Multitasking in the Human Brain." Nature Communications 16, article no. 3057 (March 29, 2025). https://doi.org/10.1038/s41467-025-58228-0.

## 1.4 The observer within

[1] Chalmers, David J. "Facing Up to the Problem of Consciousness." *Journal of Consciousness Studies* 2, no. 3 (1995): 200–219.

[2] Financial Times. "Sleep Aids to Make Your Dreams Electric." *Financial Times*, 2024.

## 1.5 Brain and Consciousness

[1] Wairagkar, Maitreyee, Nicholas S. Card, Tyler Singer-Clark, Xianda Hou, Carrina Iacobacci, Lee M. Miller, Leigh R. Hochberg, David M. Brandman, and Sergey D. Stavisky. "An Instantaneous Voice-Synthesis Neuroprosthesis." Nature 644 (2025): 145–152. https://doi.org/10.1038/s41586-025-09127-3.

[2] Cernera, S., et al. "Sustained Clinical Benefit of Adaptive Deep Brain Stimulation in Parkinson's Disease Using Gamma Oscillations: A Case Report." Movement Disorders, 2025. Wang, N., et al. "Exploring Creativity in Human–AI Co-Creation." Frontiers, September 29, 2025.

[3] Kalantari, Fatemeh, Karim Faez, Hamidreza Amindavar, and Soheila Nazari. "Improved Image Reconstruction from Brain Activity through Automatic Image Captioning." Scientific Reports 15 (2025): Article 4907. https://doi.org/10.1038/s41598-025-89242-3

## 1.6 Being Human

[1] Wang, N., H. Kim, J. Peng, and J. Wang. "Exploring Creativity in Human–AI Co-Creation: A Comparative Study across Design Experience." *Frontiers in Computer Science* 7 (2025): 1672735. https://doi.org/10.3389/fcomp.2025.1672735

[2] Lee, Hao-Ping, Advait Sarkar, Lev Tankelevitch, Ian Drosos, Sean Rintel, Richard Banks, and Nicholas Wilson. "The Impact of Generative AI on Critical Thinking: Self-Reported Reductions in Cognitive Effort and Confidence Effects from a Survey of Knowledge Workers." Proceedings of the 2025 CHI Conference on Human Factors in Computing Systems, 2025. Life Biosciences. "First Human Cellular Reprogramming Trial." Lifespan Research Institute, February 6, 2026.

[3] Lee, Hao-Ping, Advait Sarkar, Lev Tankelevitch, Ian Drosos, Sean Rintel, Richard Banks, and Nicholas Wilson. "The Impact of Generative AI on Critical Thinking: Self-Reported Reductions in Cognitive Effort and Confidence Effects From a Survey of Knowledge Workers." In Proceedings of the 2025 CHI Conference on Human Factors in Computing Systems. New York: Association for Computing Machinery, 2025.

## 1.7 The oldest temptation

[1] "FDA Go-Ahead to Test Cellular Rejuvenation Therapy in Humans." Nature Biotechnology 44 (2026): 164. https://doi.org/10.1038/s41587-026-03037-z

[2] Trouillet, Alix, Emilie Revol, Florent-Valéry Coen, et al. "High-Resolution Prosthetic Hearing with a Soft Auditory Brainstem Implant in Macaques." Nature Biomedical Engineering 9, no. 9 (2025): 1403–17. https://doi.org/10.1038/s41551-025-01378-9

[3] Mullin, Emily. "This Beanie Is Designed to Read Your Thoughts." Wired, April 16, 2026.

## 2.1 The birth of the upgrade dream

[1] Huxley, Julian. "Transhumanism." In *New Bottles for New Wine*, 13–17. New York: Harper & Brothers, 1957.

## 2.2 Crops to Genes

[1] Cyranoski, David, and Heidi Ledford. "Genome-Edited Baby Claim Provokes International Outcry." *Nature* 563 (2018): 607–8. https://doi.org/10.1038/d41586-018-07545-0

[2] Greely, H. T. "CRISPR'd Babies: Human Germline Genome Editing." 2019.

## 2.3 The transhumanist vision

[1] Reuters. "Elon Musk's Neuralink Implants Brain Chip in First Human." Reuters, January 29, 2024.

[2] Levy, Steven. "Synchron, a Rival to Musk's Neuralink, Readies Large-Scale Brain Implant Trial." *Reuters*, April 8, 2024.

[3] Endovascular Today. "Synchron's Brain-Computer Interface Evaluated in COMMAND Early Feasibility Study." Endovascular Today, October 3, 2024.

[4] Paradromics. "Paradromics Completes First-In-Human Recording." Paradromics, June 2, 2025.

[5] Reuters. "Neuralink's Speech Restoration Device Gets FDA's 'Breakthrough' Tag." Reuters, May 1, 2025.

## 2.4 Assistance, repair or enhancement

1 Ronald Cole-Turner, ed., Transhumanism and the Body: The World Religions Speak (Palgrave Macmillan, 2011).

2 Brent Waters, This Mortal Flesh: Incarnation and Bioethics (Grand Rapids: Brazos Press, 2009), quoted in Theology Forum, February 11, 2010.

## 3.1 Artificial Intelligence 101

1 Rosenblatt, Frank. "The Perceptron: A Probabilistic Model for Information Storage and Organization in the Brain." Psychological Review 65, no. 6 (1958): 386–408.

2 Weizenbaum, Joseph. "ELIZA—A Computer Program for the Study of Natural Language Communication Between Man and Machine." Communications of the ACM 9, no. 1 (January 1966): 36–45.

## 3.2 The Real AI

1 Reuters. "Trouble with AI 'Hallucinations' Spreads to Big Law Firms." Reuters, May 23, 2025.

2 Federal Court of Australia. "Use of Generative Artificial Intelligence Practice Note (GPN-AI)." April 16, 2026. https://www.fedcourt.gov.au/law-and-practice/practice-documents/practice-notes/gpn-ai.

3 Reuters. "AI Deepfakes Blur Reality in 2026 U.S. Midterm Campaigns." Reuters, March 28, 2026

## 3.3 The Illusion of thought

1 Weizenbaum, Joseph. "ELIZA—A Computer Program for the Study of Natural Language Communication between Man and Machine." Communications of the ACM 9, no. 1 (1966): 36–45. https://doi.org/10.1145/365153.365168

## 3.5 The necessity of aliens

1 Farah, Dan, dir. *The Age of Disclosure*. Farah Films, 2025.

## 4.1 The alignment problem

1 Covington, Paul, Jay Adams, and Emre Sargin. "Deep Neural Networks for YouTube Recommendations." In *Proceedings of the 10th ACM Conference on Recommender Systems*, 191–98. New York: Association for Computing Machinery, 2016. https://doi.org/10.1145/2959100.2959190

2 Reuters. "UK Regulators Rush to Assess Risks of Latest Anthropic AI Model, FT Reports." Reuters. April 12, 2026.

## 4.2 Tool vs authority


1 Bureau d'Enquêtes et d'Analyses pour la Sécurité de l'Aviation Civile. Final Report: On the Accident on 1 June 2009 to the Airbus A330-203 Registered F-GZCP Operated by Air France Flight AF 447 Rio de Janeiro–Paris. Le Bourget, France: BEA, 2012.

2 Joseph Ax and Helen Coster, "AI Deepfakes Blur Reality in 2026 US Midterm Campaigns," Reuters, March 28, 2026.


## 4.3 The brain-computer interfaces


1 Neuralink. "Neuralink's First-in-Human Clinical Trial Is Open for Recruitment." September 19, 2023. https://neuralink.com/updates/first-clinical-trial-open-for-recruitment/

2 ClinicalTrials.gov. "Precise Robotically IMplanted Brain-Computer InterfacE (PRIME Study)." NCT06429735. U.S. National Library of Medicine. https://clinicaltrials.gov/study/NCT06429735

3 Synchron. "Synchron to Begin Patient Enrollment for the COMMAND Trial of Brain-Computer Interface at Gates Vascular Institute." Business Wire, March 20, 2023. https://www.businesswire.com/news/home/20230320005156/en/Synchron-to-Begin-Patient-Enrollment-for-the-COMMAND-Trial-of-Brain-Computer-Interface-at-Gates-Vascular-Institute.

4 Mitchell, Peter, Sarah C. M. Lee, Peter E. Yoo, Andrew Morokoff, Rahul P. Sharma, Daryl L. Williams, Christopher MacIsaac, et al. "Assessment of Safety of a Fully Implanted Endovascular Brain-Computer Interface for Severe Paralysis in 4 Patients: The Stentrode With Thought-Controlled Digital Switch (SWITCH) Study." JAMA Neurology 80, no. 3 (2023): 270–278. https://doi.org/10.1001/jamaneurol.2022.4847.

5 Kirton, Adam. "A Moral Imperative to Advance Brain-Computer Interfaces for Children With Neurological Disability." JAMA Pediatrics 177, no. 8 (2023): 751–752. https://doi.org/10.1001/jamapediatrics.2023.1744. Blanke, Olaf, Theodor Landis, Laurent Spinelli, and Margitta Seeck. "Stimulating Illusory Own-Body Perceptions." Nature 443 (2006): 287. https://doi.org/10.1038/443287a.

6 European Parliament. "Resolution of 16 February 2017 with Recommendations to the Commission on Civil Law Rules on Robotics (2015/2103(INL))." Official Journal of the European Union, February 16, 2017. https://www.europarl.europa.eu/doceo/document/TA-8-2017-0051_EN.html

7 Reuters. "Saudi Women Riled by Robot with No Hijab and More Rights than Them." *Reuters*, November 1, 2017.

8 United Nations Development Programme. "Sophia the Robot Takes Part in the Fifth OGP Global Summit in Georgia." July 18, 2018. https://www.undp.org/georgia/press-releases/sophia-robot-takes-part-fifth-ogp-global-summit-georgia

9 Martínez, E., and C. Winter. "Protecting Sentient Artificial Intelligence: A

Survey of Lay Intuitions on Standing, Personhood, and General Legal Protection." Frontiers in Robotics and AI 8 (2021): Article 788355. https://doi.org/10.3389/frobt.2021.788355.

10 Pew Research Center. "2023–24 U.S. Religious Landscape Study Interactive Database." 2025. https://doi.org/10.58094/3zs9-jc14

## 4.4 Post human world

1 Human Rights Watch. "China's Chilling 'Social Credit' Blacklist." December 12, 2017. hrw.org/news/2017/12/12/chinas-chilling-social-credit-blacklist.

2 European Parliament and Council of the European Union. Regulation (EU) 2024/1689 of 13 June 2024 Laying Down Harmonised Rules on Artificial Intelligence (Artificial Intelligence Act). Official Journal of the European Union, July 12, 2024. https://eur-lex.europa.eu/eli/reg/2024/1689/oj/eng

3 Organisation for Economic Co-operation and Development. Job Creation and Local Economic Development 2024: The Geography of Generative AI. OECD Publishing, 2024. https://doi.org/10.1787/83325127-en

4 Organisation for Economic Co-operation and Development. Compendium of Best Practices for a Human-Centered Development and Use of Artificial Intelligence in the World of Work. OECD, 2025.

## 4.5 CRISPR and AI: genetic scissors to edit life

1 National Human Genome Research Institute. "CRISPR." Genome.gov. National Institutes of Health. Accessed August 15, 2026. https://www.genome.gov/genetics-glossary/CRISPR

2 Liang, Puping, Yanwen Xu, Xiya Zhang, Chenhui Ding, Rui Huang, et al. "CRISPR/Cas9-Mediated Gene Editing in Human Tripronuclear Zygotes." *Protein & Cell* 6, no. 5 (2015): 363–72. https://doi.org/10.1007/s13238-015-0153-5

3 Sibbald, Barbara. "Death but One Unintended Consequence of Gene-Therapy Trial." Canadian Medical Association Journal 164, no. 11 (2001): 1612.

4 UNESCO. Universal Declaration on the Human Genome and Human Rights. Adopted by the General Conference of UNESCO, November 11, 1997, and endorsed by the United Nations General Assembly, December 9, 1998.

5 World Health Organization. Human Genome Editing: A Framework for Governance. Geneva: World Health Organization, 2021. https://www.who.int/publications/i/item/9789240030060

6 Deborah Haarsma, "Francis Collins Speaks on Genetic Engineering and Christian Faith," BioLogos, June 18, 2018.

7 Nature. "First CRISPR Babies: Six Questions That Remain." Nature, November 30, 2018.

[8] Olawade, David B., et al. "Artificial Intelligence in In-Vitro Fertilization (IVF): A New Era of Assisted Reproductive Technology." *Reproductive Sciences* (2025)

[9] American Society for Reproductive Medicine. "Indications and Management of Preimplantation Genetic Testing for Monogenic Conditions: A Committee Opinion." 2023. https://www.asrm.org/practice-guidance/practice-committee-documents/indications-and-management-of-preimplantation-genetic-testing-for-monogenic-conditions-a-committee-opinion-2023/

[10] Glenn Cohen. "The He Jiankui Affair." Journal of Law and the Biosciences 6, no. 1 (2019): 111–183. https://doi.org/10.1093/jlb/lsz010.

[11] Stephen Chen. "Chinese Team Behind Extreme Animal Gene Experiment Says It May Lead to Super Soldiers Who Survive Nuclear Fallout." South China Morning Post, March 29, 2023. https://www.scmp.com/news/china/science/article/3215286.

[12] US Food and Drug Administration. "FDA Approves First Gene Therapies to Treat Patients with Sickle Cell Disease." Press announcement, December 8, 2023. https://www.fda.gov/news-events/press-announcements/fda-approves-first-gene-therapies-treat-patients-sickle-cell-disease.

[13] Guido Pennings, "Refining the Ethics of Preimplantation Genetic Diagnosis: A Plea for Contextualized Proportionality," Human Reproduction 33, no. 10 (2018): 1756–1763. https://doi.org/10.1093/humrep/dey285.

## 5.1 Intelligence and the Creator

[1] Science Daily. "Why Scientists Are Racing to Define Consciousness." ScienceDaily, February 1, 2026.

## 5.2 Why humans matter

[1] Gillibrand, Kirsten, and Rick Scott. Letter to the Chairman, Senate Special Committee on Aging, 2026.

[2] Microsoft. "Seeing AI." Microsoft. https://www.microsoft.com/en-us/ai/seeing-ai.

## 5.3 Technology as servant not saviour

[1] Federal Trade Commission. "FTC Launches Inquiry into AI Chatbots Acting as Companions." FTC.gov, September 2025.

www.ingramcontent.com/pod-product-compliance
Lightning Source LLC
LaVergne TN
LVHW050801061126
851857LV00012B/630